Popular Furniture of the 1920s and 1930s

From Traditional to Early Modern

4880 Lower Valley Road, Atglen, PA 19310 USA

Published by Schiffer Publishing Ltd.
4880 Lower Valley Road
Atglen, PA 19310
Phone: (610) 593-1777; Fax: (610) 593-2002
e-mail: schifferbk@aol.com
Please write for a free catalog.
This book may be purchased from the publisher.
Please include $3.95 for shipping.

Please try your bookstore first.

We are interested in hearing from authors
with book ideas on related subjects.

Copyright © 1998 by Schiffer Publishing Ltd.
Library of Congress Catalog Card Number: 97-80769

All rights reserved. No part of this work may be
reproduced or used in any form or by any
means—graphic, electronic, or mechanical, including
photocopying or information storage and retrieval
systems—without written permission from the copyright
holder.

ISBN: 0-7643-0431-3
Printed in the United States of America

Introduction

Popular Furniture of the 1920s and 1930s is a facsimile reproduction of the Elgin A. Simonds Company's furniture catalog. It presents the reader with an extensive resource of traditional and commercial furniture styles of the 1920s and 1930s.

The Elgin A. Simonds Company was part of a consortium of furniture manufacturers which also included The Lenox Shops Inc. of Canastota, New York and The Lincoln Furniture Company of Philadelphia, Pennsylvania. These three companies where identified under the umbrella of The Dent Furniture Corporation. By combining their diverse product offerings, The Dent Furniture Corporation was able to offer both the residential and commercial buyer a comprehensive line of stylish and functional furniture.

Founded in 1901, The Elgin A. Simonds Company engaged exclusively in the production of faithful reproductions of the finest work of master furniture craftsmen of the past. The company president would spend much of his time researching styles in Europe looking for patterns that were worthy of reproduction by the craftsmen at the Simonds Company.

Current values of the furniture have been added to make this book a useful addition to the collector's library.

I would have our dwelling houses built to be lovely, as rich and full of pleasantries as may be, within and without.

Ruskin: "The Seven Lamps of Architecture."

Simonds Quality Products

FOR over a quarter of a century The Elgin A. Simonds Company has engaged exclusively in producing faithful reproductions of the best work of the master furniture craftsmen of the past. The President of this Company devotes many months each year to searching in Europe for patterns worthy of reproduction.

Quality has been the keynote from the very beginning. The Elgin A. Simonds Quality is recognized and accepted by both the dealer and user as the standard of excellence. Home owners feel a pride of possession through the purchase of The Elgin A. Simonds Individualized Good Furniture that no other inspires nor can satisfy so well.

Dealers enjoy a prestige for quality that the sale of ordinary lines cannot give. There is artistry and craftsmanship reflected in every piece of Elgin A. Simonds Furniture. It is the line that stands out from competition, thereby reducing sales resistance to a minimum. The dealer who handles Elgin A. Simonds Furniture will unquestionably build a profitable and permanent business.

THE ELGIN A. SIMONDS COMPANY
SYRACUSE, NEW YORK

PERMANENT SHOWROOMS

NEW YORK	GRAND RAPIDS	PHILADELPHIA
101 Park Avenue	Klingman-Waters Building	Lincoln Furniture Co.
322 North Wing		19th and Lehigh

CHICAGO
645 St. Clair Street, Corner Erie Street

INDIVIDUALISM
In Good Furniture

as produced by three quality manufacturers affiliated with

The Dent Furniture Corporation

THE ELGIN A. SIMONDS COMPANY
Syracuse, N. Y.
Founded 1901

THE LINCOLN FURN. COMPANY
Philadelphia, Pa.
Founded 1853

THE LENOX SHOPS, INC.
Canastota, N. Y.
Founded 1893

INDIVI
In Go

As produced

THE DENT FUI

This correct group harmo
operation of skillful desi
the one goal of produc
in Good Furniture. I
of manufacture that
Furniture Corporati
hold in the Fine I
The extent of th
makes it possib
zation practic
ho

THE EL

THE

T

An attractive interior above, shows an attractive arrangement of furniture from the Elgin A. Simonds Company. The selection presents an effect that is pleasingly individual. The view below presents a spacious Club Lounge using a combination of Lenox and Simonds pieces in characteristic grouping.

The bedroom below, from the Half Moon Hotel, is furnished by the Lincoln Furniture Company. These well chosen pieces lend to an otherwise unattractive room that home like atmosphere that is so desirable. The chair in this group is by Simonds and takes its place gracefully with the other pieces.

These three plants offer a complete service to the dealer by covering thoroughly the field of distinct-home furnishings. The designs are made to appeal to that ever increasing majority of the public who have long recognized fidelity of design, fine workmanship and honest materials as constituting true value. All executives of The Dent Furniture Corporation are ever alert to the changing modes in the production of fine furniture. The present mode has been fostered and pieces developed that are in harmony with it. You will find in the catalogs groups and individual pieces which have entirely new motifs —designed in the modern manner.

JALISM
niture
manufacturers

CORPORATION

ssible through the close co-
e before them continually
at bespeak Individualism
ualism and the honesty
roducts of The Dent
d place that they now
p of Manufacturers.
es of these plants
one great organi-
needed for the
tel.

Company

OMPANY

Inc.

The Dining Room group above is by Lenox and shows a modern adaptation of an authentic Sheraton design. Each individual piece is beautifully proportioned and in combination with the other pieces produces an interior that is interesting and appeals to the discriminating purchaser. Many other attractive Dining Room groups are displayed in the Lenox line.

The Office group shown above is a combination of Simonds and Lenox pieces that blend well in giving to the office an atmosphere of hospitality and security. The spacious living room below presents another grouping of pieces from these same two plants, producing an atmosphere of charm and individuality.

Constructional features conform with the highest standards of the industry. All exposed parts are of the wood designated for the different numbers in price list or catalog; case work is built with three-ply dust panels; assembly is made for the life-time rigidity and service required of furniture that will always be in vogue. If your files do not include Simonds, Lincoln and Lenox catalogs you will confer a favor by notifying any one of the three plants. These three catalogs, covering completely the field of fine furniture making, constitute a work of reference on the present day tastes for Individualism in Good Furniture.

*Permanent Showrooms
where Simonds, Lenox and Lincoln Furniture
is on display at all times*

GRAND RAPIDS
PERMANENT SHOWROOM
Klingman-Waters Building

LOS ANGELES
PERMANENT SHOWROOM
816 South Figueroa Street

CHICAGO
PERMANENT SHOWROOM
645 St. Clair Street, Corner of Erie

NEW YORK CITY
PERMANENT SHOWROOM
101 Park Avenue, 332 North Wing

Contents

Pages

American Colonial	10, 11, 13, 14, 16, 22, 23, 27, 28, 52, 55, 56, 71, 73, 76, 77, 98, 113, 114, 115, 125, 136, 137, 138, 142, 146, 148, 151, 157, 159, 162, 163, 165, 169
Bed Room Chairs and Sets	58, 73, 110, 112, 113, 114, 115, 116, 117, 118
Benches	131, 132, 133, 134, 135
Bookstands	141, 144, 145, 147, 155, 158, 163
Breakfast and Dining Room Sets	44, 45, 135, 136, 137, 138, 170, 171
Card Chairs and Tables	29, 51, 57, 66, 110, 112, 129, 147
Cellarettes	145, 149
Chaise Longue	80, 81
Coffee Tables, Stands and Nest Tables	136, 139, 140, 141, 142, 143, 144, 145, 146, 147, 148, 149, 150, 153, 155
Consoles and Mirrors	164, 165, 166, 167, 168, 169
Coxwell Chairs	68, 69, 92, 95, 103
Desks and Cabinets	162, 163, 164
Desk Chairs	4, 6, 7, 8, 9
Dining Chairs	26, 27, 28, 31, 32, 33, 34, 35, 36, 37, 38, 39, 40, 41, 42, 43, 44, 45, 46, 57
Early English Seating Furniture	16, 20, 28, 44, 45, 49, 65, 66, 73, 78, 105, 109, 126, 129, 130, 131, 132, 134
Early English Tables, and Casework	116, 117, 118, 128, 129, 139, 140, 141, 142, 147, 149, 151, 153, 154, 155, 156, 160, 161
Early French, Italian and Spanish Furniture	13, 16, 18, 19, 20, 24, 31, 61, 64, 74, 75, 76, 78, 89, 100, 105, 133, 152, 153, 160, 163, 167
Easy Chairs	67, 68, 69, 76, 77, 83, 84, 85, 86, 87, 88, 89, 92, 93, 94, 95, 96, 103, 105, 106, 107, 108, 109
Empire Furniture	26, 50, 54, 62, 64, 89, 107, 108, 110, 120, 132, 150, 167
End Tables	139, 140, 141, 142, 143, 146, 149, 151
Gate Leg Tables	136, 156
Grill Chairs	10, 11, 27, 31
High Upholstered Back Chairs	48, 49, 56, 60, 61, 62, 74, 75, 76, 77
High Wood Back Chairs	15, 16, 17, 18, 19, 20, 21, 44, 66
Large Tables	151, 152, 153, 154, 155, 156, 157, 158, 159, 160, 161, 165
Late French Furniture	50, 53, 62, 63, 64, 68, 72, 81, 112, 119, 122, 124, 132, 140, 142, 144, 166
Love Seats	75, 79, 80, 81, 83, 92, 94, 97, 100, 106, 111, 114, 120, 124
Moderne	110, 111, 112, 128, 141, 155, 171
Occasional Chairs	47, 48, 49, 50, 51, 52, 53, 54, 55, 56, 57, 58, 59, 60, 61, 62, 63, 64, 65, 66, 77, 78, 79, 80, 86, 87, 88, 89, 97, 98, 99, 100, 104, 110, 111, 112, 113
Office Desks and Chairs	9, 28, 29, 57, 61, 124, 125, 126, 127, 128
Ottomans and Foot Stools	57, 67, 68, 69, 130
Screens	45
Smoker Stands	139, 140, 141, 142, 151
Sofas and Settees	12, 15, 19, 22, 26, 66, 76, 77, 78, 82, 83, 84, 85, 86, 87, 88, 89, 90, 91, 92, 93, 94, 95, 96, 98, 99, 100, 101, 102, 103, 104, 105, 106, 107, 108, 109, 114, 120, 121, 122, 123, 124, 134
Solarium Furniture	110, 111, 112, 113, 114, 115, 144
Vanities and Chairs	26, 118, 119, 124
Windsors	5, 6, 7, 8, 9, 10, 11, 12
Wing Chairs	62, 70, 71, 72, 73, 75, 76, 81, 82, 84, 85, 88, 92, 93, 96, 108, 114
LENOX PATTERNS	173 to 180
LINCOLN PATTERNS	181 to 219

Windsor and Desk Chairs

$145

2169 SC & AC—Simplified Windsors of period about 1725. Selected birch; inexpensive; suitable for breakfast rooms. Antique finish.

$165

2169 SC
Height 37 inches
Across the Seat 17 inches
Depth of Seat 15 inches
Code Word—BESCA

2169 AC
Height 37 inches
Across the Seat 20 inches
Depth of Seat 16 inches
Code Word—BEZAC

$185

993 AC—An English Windsor, two-tier back with Chippendale splat. Antique finish.

996 SC—A braced-back Windsor with Chippendale splat. Antique finish.

$125

993 AC
Height 37 inches
Across the Seat 20 inches
Depth of Seat 16 inches
Code Word—ACHIS

996 SC
Height 36 inches
Across the Seat 17 inches
Depth of Seat 15 inches
Code Word—ACIPA

Windsor Chairs

$155

1579 SC & AC—Windsors of the period 1725–1750. Particularly comfortable and attractive. Antique finish. Hand flagged seat.

$210

1579 SC
Height 36 inches
Across the Seat 16 inches
Depth of Seat 15 inches
Code Word—AKIFY

1579 AC
Height 38 inches
Across the Seat 21½ inches
Depth of Seat 15 inches
Code Word—AKILE

$145

979 SC—Braced Bow-back Windsor, same model as original in Iron Works House, Lauzus Centre, Mass.

1858 AC—A popular type in American homes between 1750 and 1800. Antique finish.

$220

979 SC
Height 36 inches
Across the Seat 16 inches
Depth of Seat 16 inches
Code Word—ACCIM

1858 AC
Height 38 inches
Across the Seat 21½ inches
Depth of Seat 16 inches
Code Word—PUFOY

SIMONDS

Windsor and Desk Chairs

$135

1884 SC & AC—Strong and substantial in construction. Note the back braces. Models came from Herts County, England.

$165

1884 SC
Height 35 inches
Across the Seat 17 inches
Depth of Seat 15 inches
Code Word—PEYSE

1884 AC
Height 37 inches
Across the Seat 20 inches
Depth of Seat 16 inches
Code Word—PEYUG

$210

1890 AC—An unusual type of English Windsor chair. Model from Victoria and Albert Museum.

1886 AC—The original of this chair came from England and is made of Yew, the shapes in the back and arms being the natural curves, in the Yew.

$225

1890 AC
Height 42 inches
Across the Seat 24 inches
Depth of Seat 16 inches
Code Word—PEYWI

1886 AC
Height 47 inches
Across the Seat 23½ inches
Depth of Seat 17 inches
Code Word—PEZAN

Windsor and Desk Chairs

$210

1882 AC—An extremely graceful type of English Windsor, Queen Anne style.

1874 AC—An English Windsor unusual in size and style having Queen Anne front legs and Gothic motives in the splats.

$220

1882 AC
Height 42 inches
Across the Seat 21½ inches
Depth of Seat 16 inches
Code Word—PEZBO

1874 AC
Height 33 inches
Across the Seat 20 inches
Depth of Seat 14 inches
Code Word—PEZES

$180

1880 AC—Model of this English Windsor chair was bought in High Wycombe.

1872 AC—A farmhouse type of English Windsor chair. Model found in High Wycombe.

$170

1880 AC
Height 43 inches
Across the Seat 18½ inches
Depth of Seat 16 inches
Code Word—PEZGU

1872 AC
Height 39 inches
Across the Seat 19 inches
Depth of Seat 16 inches
Code Word—PEZGU

Windsor and Desk Chairs

$135

1802 SC & AC—Old English coffee-house model with bell leg and Gothic arched back. Antique finish.

$180

1802 SC
Height 41 inches
Across the Seat 20 inches
Depth of Seat 16 inches
Code Word—ADABE

1802 AC
Height 43 inches
Across the Seat 22 inches
Depth of Seat 16 inches
Code Word—ADAFI

$175

967 AC—A later development of the Hogarth chair, showing the Windsor influence in its spindles. The original dates from 1725. Antique finish.

1832 AC—Original Hogarth type.

$175

967 AC
Height 44 inches
Across the Seat 23 inches
Depth of Seat 17 inches
Code Word—ACBOS

1832 AC
Height 39 inches
Across the Seat 20½ inches
Depth of Seat 17 inches
Code Word—OPADE

Windsor and Desk Chairs

$175

1864 AC—Beautifully proportioned Windsor with Chippendale pierced back.

1834 AC—English Windsor of rigorous and sturdy proportions.

$210

1864 AC
Height 40 inches
Across the Seat 21½ inches
Depth of Seat 15 inches
Code Word—PUFUD

1834 AC
Height 45 inches
Across the Seat 19 inches
Depth of Seat 19 inches
Code Word—PRIKU

$225

981 AC—The "Thomas Jefferson," reproduced from the original made for and used by that American statesman. Antique finish.

1824½ Swivel AC—Swivel desk chair, developed for office or home from Oliver Goldsmith chair shown below.

$175

981 AC
Height 44 inches
Across the Seat 21 inches
Depth of Seat 16 inches
Code Word—ACEYE

1824½ Swivel AC
Height of Chair 37 inches
Across the Seat 18 inches
Depth of Seat 17 inches
Code Word—OPAGH

SIMONDS

Early American Chairs

$155

1803 AC
Height 30 inches
Across the Seat 22 inches
Depth of Seat 17 inches
Code Word—ADALO

$160

1803 AC—Excellent for office or club use.

1803½ AC—Has back, arms and seat covered with leather.

1842 AC—American low back Windsor of Pennsylvania type.

1842 AC
Height 32 inches
Across the Seat 23 inches
Depth of Seat 19 inches
Code Word—PRIEN

$180

1840 AC—Windsor arm chair. American low comb back design, as fine an example as we know of.

3800 AC—Cherry Valley Chair. The original of this quaint tavern chair was found in the historic Cherry Valley of New York State. Our reproduction of this chair has all the quaint charm of the original besides having a surprising seating comfort and ruggedness of construction.

$170

1840 AC
Height 34 inches
Across the Seat 21½ inches
Depth of Seat 16 inches
Code Word—PREZE

3800 AC
Height 29 inches
Across the Seat 19½ inches
Depth of Seat 16 inches
Code Word—PEMUA

English Models

1892 AC—A tavern chair from old England.

1889 AC—A light compact roomy pickup chair.

$150

$155

1892 AC
Height 28 inches
Across the Seat 22½ inches
Depth of Seat 15 inches
Code Word—PERZE

1889 AC
Height 31 inches
Across the Seat 19 inches
Depth of Seat 16 inches
Code Word—PESYE

$225

$120

1877 AC—A fine light weight round-about chair. Model found in Ipswich, England.

3357 AC—A reproduction of the Old Colonial corner chair.

1877 AC
Height 30 inches
Across the Seat 23 inches
Depth of Seat 22 inches
Code Word—PEVUD

3357 AC
Height 35 inches
Across the Seat 24 inches
Depth of Seat 18 inches
Code Word—PEVZI

Windsor Chairs

$175

1812 AC—A designed chair of Windsor type.

1836 AC—Queen Anne Windsor of great beauty and real comfort. Very unusual is the large carved and shaped back splat.

$185

1812 AC
Height 37 inches
Across the Seat 19½ inches
Depth of Seat 17 inches
Code Word—ADDYE

1836 AC
Height 39 inches
Across the Seat 22 inches
Depth of Seat 15 inches
Code Word—PRIFO

$185

$290

1850 AC
Height 43 inches
Across the Seat 23½ inches
Depth of Seat 17 inches
Code Word—PRIRA

1850 Settee
Height 41 inches
Overall 42 inches
Depth of Seat 17 inches
Code Word—PRIUD

1850 AC & Settee—Very unusual English Windsor with shaped splat and carved top rail Queen Anne style.

Hall and Library Chairs

2782 SC & AC—From an Italian model.

2782 SC
Height 39 inches
Across the Seat 19 inches
Depth of Seat 16 inches
Genuine Hand Flagged
Code Word—ORACE

2782 AC
Height 40 inches
Across the Seat 22½ inches
Depth of Seat 19 inches
Genuine Hand Flagged
Code Word—ORAEG

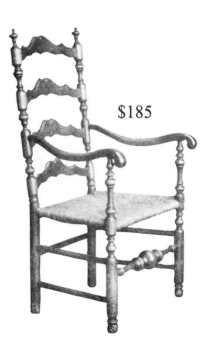

2502 SC & AC—"The Norwich"—solid mahogany, hand-made flag seat. Period 1790.

2502 SC
Height 43 inches
Across the Seat 19 inches
Depth of Seat 17 inches
Genuine Hand Flagged
Code Word—BIBEM

2502 AC
Height 45 inches
Across the Seat 24 inches
Depth of Seat 19 inches
Genuine Hand Flagged
Code Word—BOBOX

SIMONDS

Hall and Library Chairs

$210

735 SC & AC—The original of this type was a splendid example of the fiddle-back pattern, popular during the early period of the reign of Queen Anne; about 1715. Hand flagged seats.

$285

735 SC
Height 42 inches
Across the Seat 19 inches
Depth of Seat 14 inches
Code Word—AGSCA

735 AC
Height 43 inches
Across the Seat 22 inches
Depth of Seat 16 inches
Code Word—AGACI

$185

2541 AC & SC—"The Albany"—An unusual and beautiful type of chair dating from the Pilgrim century. Hand flagged seats.

$210

2541 SC
Height 41 inches
Across the Seat 19 inches
Depth of Seat 17 inches
Code Word—BILAT

2541 AC
Height 44 inches
Across the Seat 23 inches
Depth of Seat 19 inches
Code Word—BIMAU

SIMONDS

Living Room Furniture

2972 SC & AC—Magnificent Carolean Suite, heavily carved. The caning is very fine.

$210

$285

2972 SC
Height 42 inches
Across the Seat 19 inches
Depth of Seat 17 inches
Code Word—BUOKS

2972 AC
Height 46 inches
Across the Seat 24 inches
Depth of Seat 19 inches
Code Word—BUOTA

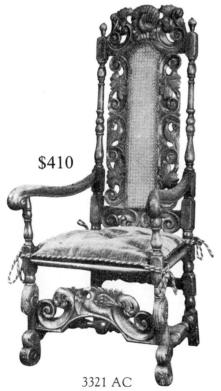

$410

$500

3321 AC
Height 51 inches
Across the Seat 23 inches
Depth of Seat 17 inches
Code Word—PEDCU

2972 Settee
Height 47 inches
Length 51 inches
Depth of Seat 19 inches
Code Word—BUPEM

2972½ AC, SC & Settee—Like 2972 but made with upholstered seat and back.

AC—Code Word—PUGJU
SC—Code Word—PUGIT
Settee—Code Word—PUGNY

3321 AC—This formal Jacobean chair would adorn any man's home. The carving is especially fine in modelling and detail.

Hall and Library Chairs

$225

2964 SC & AC—The carved inset medallion and reversed twist pilasters of this Charles II chair are particularly attractive. Hand caned seat and back.

$365

2964 SC
Height 44 inches
Across the Seat 19 inches
Depth of Seat 16 inches
Code Word—BUNAG

2964 AC
Height 47 inches
Across the Seat 24 inches
Depth of Seat 18 inches
Code Word—BUNEK

$220

1697 AC & SC—Early English-Spanish influence straight back arm and side chairs belonging to the 18th Century. Antique finish.

$200

1697 AC
Height 52 inches
Across the Seat 23½ inches
Depth of Seat 17 inches
Code Word—ALMEC

1697 SC
Height 45 inches
Across the Seat 19 inches
Depth of Seat 15 inches
Code Word—ALMAY

SIMONDS

Hall and Library Chairs

$225

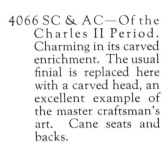

4066 SC & AC—Of the Charles II Period. Charming in its carved enrichment. The usual finial is replaced here with a carved head, an excellent example of the master craftsman's art. Cane seats and backs.

$410

4066 SC
Height 51 inches
Across the Seat 19 inches
Depth of Seat 15 inches
Code Word—FOPUT

4066 AC
Height 54 inches
Across the Seat 23½ inches
Depth of Seat 17 inches
Code Word—FOPYX

$230

1790 SC & AC—Charles II chairs of great distinction.

$395

1790 SC
Height 57 inches
Across the Seat 23 inches
Depth of Seat 17 inches
Code Word—AMTUZ

1790 AC
Height 62 inches
Across the Seat 26 inches
Depth of Seat 19 inches
Code Word—AMTYD

SIMONDS

Living Room Furniture

2952 AC & SC—Italian XVII Century.

$210

$335

2952 SC
Height 42 inches
Across the Seat 19 inches
Depth of Seat 17 inches
Code Word—OSABE

2952 AC
Height 47 inches
Across the Seat 25 inches
Depth of Seat 21 inches
Code Word—OSAFI

4062 AC & SC—Italian 16th Century. Chairs very much like these are treasured in the Florence National Museum.

$310

$210

4062 AC
Height 54 inches
Across the Seat 24 inches
Depth of Seat 20 inches
Code Word—PUVBA

4062 SC
Height 48 inches
Across the Seat 20 inches
Depth of Seat 16 inches
Code Word—PUVED

SIMONDS

Hall and Library Chairs

2982½ SC, AC & Settee & 3030 AC—Italian XVI Century. Especially good for narrow halls and on staircase landings. Hand-carved. Antique finish.

$205

2982½ SC
Height 43 inches
Across the Seat 18 inches
Depth of Seat 14 inches
Code Word—BURUD

$310

3030 AC
Height 46 inches
Across the Seat 25 inches
Depth of Seat 17 inches
Code Word—PEPKO

$310

2982½ AC
Height 46 inches
Across the Seat 23½ inches
Depth of Seat 15 inches
Code Word—BUSAK

$550

2982½ Settee
Height 48 inches
Length 49 inches
Depth of Seat 15 inches
Code Word—BUSEO

Hall and Library Chairs

1766 AC
Height 50 inches
Across the Seat 23 inches
Depth of Seat 18 inches
Code Word—AMPYA

$295

1766 SC & AC—The Jacobean influence on Spanish furniture is illustrated in these handsome hall or library chairs.

$210

1766 SC
Height 46 inches
Across the Seat 19 inches
Depth of Seat 17 inches
Code Word—AMPOR

$295

2613 SC & AC—Carolean hall chairs with finely carved details. Cane seats.

$320

2613 SC
Height 46 inches
Across the Seat 20 inches
Depth of Seat 16 inches
Code Word—BLOON

2613 AC
Height 53 inches
Across the Seat 24 inches
Depth of Seat 16 inches
Code Word—BLOFE

Hall and Library Chairs

$285

1125½ SC & AC—Same type as 1588 chairs, with caned back panel.

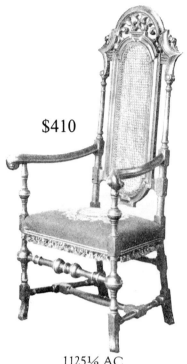

$410

1125½ SC
Height 50 inches
Across the Seat 19 inches
Depth of Seat 15 inches
Code Word—AIDAL

1125½ AC
Height 55 inches
Across the Seat 23½ inches
Depth of Seat 16 inches
Code Word—AIDDO

$300

1588 SC & AC—William & Mary hall chairs. Noteworthy for the beauty of their carving. The richness of the walnut blends into the soft tones of the velvet seat and panel. Antique finish.

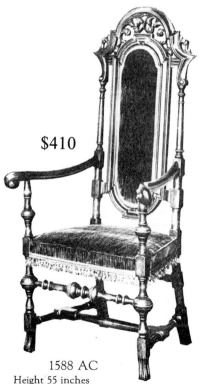

$410

1588 SC
Height 50 inches
Across the Seat 19 inches
Depth of Seat 15 inches
Code Word—AKOGE

1588 AC
Height 55 inches
Across the Seat 23½ inches
Depth of Seat 16 inches
Code Word—AKOKI

Living Room Furniture

$375

2555 Settee—Seventeenth century turned Settee. Genuine hand flagged seat. Chairs to match on page 23.

$445

1718 AC
Height 37 inches
Across the Seat 23 inches
Depth of Seat 21 inches
Genuine Hand Flagged
Code Word—AMENE

2555 Settee
Height 44 inches
Length 38 inches
Depth of Seat 17 inches
Code Word—PATUX

$550

$210

1718 Settee
Height 36 inches
Length 52 inches
Depth of Seat 21 inches
Genuine Hand Flagged
Code Word—AMESI

1718 SC
Height 35 inches
Across the Seat 20 inches
Depth of Seat 18 inches
Genuine Hand Flagged
Code Word—AMEJA

1718 AC, Settee & SC—Normandy Chairs and Settee of great charm. The low pitched back and graceful swell of the stretcher are especially pleasant. From an original in the Musee Arletan, Department of Calvados, France. Antique finish.

2899 SC & AC—Like 1718 but made with upholstered seats.
SC—Code Word—PUGAK
AC—Code Word—PUGEO

Early American Chairs

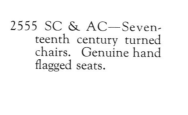

2555 SC & AC—Seventeenth century turned chairs. Genuine hand flagged seats.

$210

$350

2555 SC
Height 42 inches
Across the Seat 20 inches
Depth of Seat 15 inches
Code Word—BIRGE

2555 AC
Height 44 inches
Width overall 22 inches
Depth of Seat 17 inches
Code Word—BIROM

$345

1694 SC & AC—"The William Penn"—simple yet very attractive; with hand flagged seat.

These chairs were found throughout Colonial Pennsylvania and Maryland.

$210

1694 AC
Height 45 inches
Across the Seat 22½ inches
Depth of Seat 19 inches
Genuine Hand Flagged
Code Word—ALLUR

1694 SC
Height 43 inches
Across the Seat 19 inches
Depth of Seat 17 inches
Genuine Hand Flagged
Code Word—ALLIF

Painted Furniture

2621 SC & AC—Unusual type. Of Spanish origin.

$235

$375

2621 SC
Height 39 inches
Across the Seat 19 inches
Depth of Seat 16 inches
Genuine Hand Flagged
Code Word—BLYGO

2621 AC
Height 40 inches
Across the Seat 22½ inches
Depth of Seat 18 inches
Genuine Hand Flagged
Code Word—BLYOX

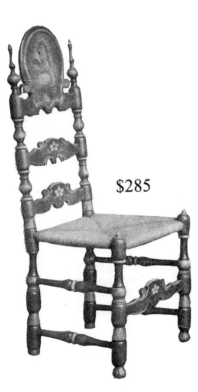

2523 SC & AC—Toledo chairs with very unusual outlines and colorful decoration from old Spanish Antique.

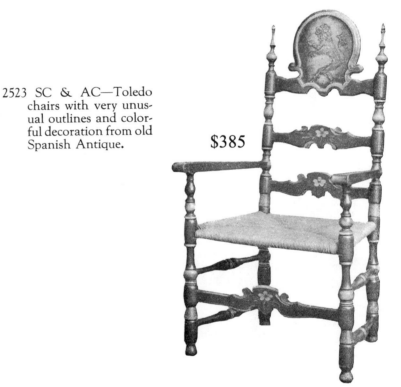

$285

$385

2523 SC
Height 42 inches
Across the Seat 19 inches
Depth of Seat 16 inches
Genuine Hand Flagged
Code Word—BIFLY

2523 AC
Height 45 inches
Across the Seat 24 inches
Depth of Seat 18 inches
Genuine Hand Flagged
Code Word—BIFNA

Painted Chairs

$210

2852 SC & AC—Sheraton decorated chairs. Having the grace, beauty and comfort of that old master's work.

$300

2852 SC
Height 34 inches
Across the Seat 20 inches
Depth of Seat 16 inches
Code Word—PRUAV

2852 AC
Height 35 inches
Across the Seat 22 inches
Depth of Seat 18 inches
Code Word—PRUDY

$210

3214 AC—Small Sheraton arm and side chairs painted and decorated.

$300

3214 SC
Height 31 inches
Across the Seat 18 inches
Depth of Seat 15 inches
Code Word—PEMJO

3214 AC
Height 32 inches
Across the Seat 19 inches
Depth of Seat 18 inches
Code Word—PUBYD

Hand Decorated Furniture

$195

3299—Painted directoire type.

$185

2852½ SC
Height 31 inches
Across the Seat 19 inches
Depth of Seat 16 inches
Code Word—PEPOT

3299 SC
Height 28 inches
Across the Seat 16 inches
Depth of Seat 15 inches
Code Word—PERAF

2852½ Suite—Sheraton in style suitable for reception or morning room. Hand painted decorations.

$295

$245

2852½ Settee
Height 32 inches
Length 40 inches
Depth of Seat 17 inches
Code Word—PEPWA

2852½ AC
Height 32 inches
Across the Seat 22 inches
Depth of Seat 17 inches
Code Word—PEPUY

Early American Chairs

$395

3067 AC—This Chippendale arm chair copied from a model found in Ipswich, England, is most unusual in its proportions. It has an extreme width of 31 inches and a height of 36 inches. The seat is low, being only 15 inches high, giving the chair a broad squatty appearance.

$210

3067 AC
Height 36 inches
Across the Seat 26 inches
Depth of Seat 19 inches
Code Word—PEMEH

2238 SC—Italian.
Height 35 inches
Across the Seat 18 inches
Depth of Seat 16 inches
Code Word—PEMDI

$395

1576 SC & AC—Pair of chairs with Spanish feet belonging to the Pilgrim Century. Cane seats and backs.

$265

1576 AC
Height 52 inches
Across the Seat 22 inches
Depth of Seat 17 inches
Code Word—OPAXY

1576 SC
Height 47 inches
Across the Seat 18 inches
Depth of Seat 15 inches
Code Word—OPAUV

SIMONDS

Living Room Furniture

$195

$345

4001 AC & SC—Cinque cento Italian Wainscot chair. Antique finish.

4001 SC
Height 39 inches
Across the Seat 19 inches
Depth of Seat 16 inches
Code Word—PUCAG

4001 AC
Height 39 inches
Across the Seat 23 inches
Depth of Seat 18 inches
Code Word—FOBAK

$175

$300

3216 AC & SC—Early type of arm chair of good proportion suitable for use in simple interiors of all kinds.

3216 SC
Height 33 inches
Across the Seat 20 inches
Depth of Seat 16 inches
Code Word—PUCCI

3216 AC
Height 34 inches
Across the Seat 22 inches
Depth of Seat 17 inches
Code Word—PUCEK

Occasional Chairs

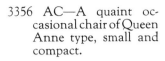

$275

3356 AC—A quaint occasional chair of Queen Anne type, small and compact.

3336 AC—The original of this chair is used in the directors' room of the Bank of England. A comfortable chair for bridge or library.

$380

3356 AC
Height 33 inches
Across the Seat 20 inches
Depth of Seat 17 inches
Code Word—PEMNO

$400

3336 AC
Height 35 inches
Across the Seat 22½ inches
Depth of Seat 17 inches
Code Word—PEMOP

30.1 AC
Height 38 inches
Across the Seat 23 inches
Depth of Seat 18 inches
Code Word—PEMYD

$365

3041 AC—A Georgian that shows a masterly sense of restraint in the use of refined classic detail. The subtle modeling of the back and the voluted finials are unusually interesting.

3361 AC—Queen Anne at its best. Perfect in design, finish and material. Model from Ipswich, England.

3341 AC—A most unusual type of Queen Anne occasional chair. Original from High Wycombe, England.

$380

3361 AC
Height 36 inches
Across the Seat 23 inches
Depth of Seat 18 inches
Code Word—PEMXY

3341 AC
Height 39 inches
Across the Seat 22½ inches
Depth of Seat 18 inches
Code Word—PEMZA

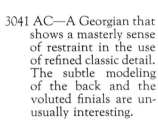

Occasional Chairs

$225

3365 AC—A modern design using Chippendale motives. Small, comfortable, and of pleasing outline.

3353 AC—Notice the French and classic influence combined with English proportions. Original probably made by Hepplewhite.

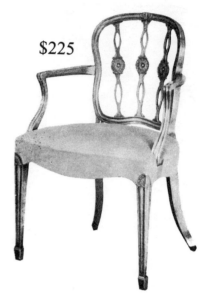

$225

3365 AC
Height 32 inches
Across the Seat 22 inches
Depth of Seat 18 inches
Code Word—PEKON

3353 AC
Height 35 inches
Across the Seat 23 inches
Depth of Seat 18 inches
Code Word—PEKPO

$300

3305 AC—Sheraton design.

3271 AC—Italian directoire painted.

$275

3305 AC
Height 32 inches
Across the Seat 20½ inches
Depth of Seat 18 inches
Code Word—PEKUT

3271 AC
Height 33 inches
Across the Seat 21½ inches
Depth of Seat 18 inches
Code Word—PEKZY

Dining and Grill Room

$275

3029 AC & SC — Small chairs, modern type.

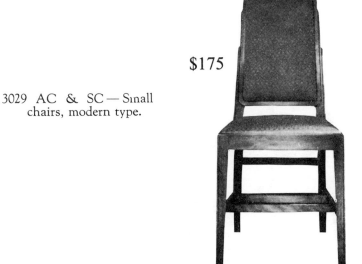

$175

3029 AC
Height 38 inches
Across the Seat 21½ inches
Depth of Seat 17 inches
Code Word—PEMIL

3029 SC
Height 37 inches
Across the Seat 18 inches
Depth of Seat 16 inches
Code Word—PEMFI

$235

1497 SC & AC—The strength and grace of Italian furniture of the 17th Century have been preserved in these modern adaptations. Companion Bench found on page 132. Stand and Library Table on page 151. Antique finish.

$245

1497 SC
Height 36 inches
Across the Seat 20 inches
Depth of Seat 17 inches
Code Word—AJOKI

1497 AC
Height 38 inches
Across the Seat 23 inches
Depth of Seat 19 inches
Code Word—AJORO

Dining Room Furniture

$280

1206 AC and SC—The graceful dignity of these Hepplewhite shield-back chairs make them particularly suitable for the more formal dining room.

$220

1206 AC
Height 39 inches
Across the Seat 22 inches
Depth of Seat 17 inches
Code Word—ABANO

1206 SC
Height 38 inches
Across the Seat 20 inches
Depth of Seat 17 inches
Code Word—ABADE

$220

3232 SC and AC—Like 1206. Has full upholstered seat.

$280

3232 SC
Height 38 inches
Across the Seat 20 inches
Depth of Seat 17 inches
Code Word—PUPIC

3232 AC
Height 39 inches
Across the Seat 22 inches
Depth of Seat 17 inches
Code Word—PUPUN

SIMONDS

Dining Room Furniture

$230

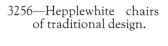

3256—Hepplewhite chairs of traditional design.

$265

3256 SC
Height 38 inches
Across the Seat 20 inches
Depth of Seat 17 inches
Code Word—PERJO

3256 AC
Height 38 inches
Across the Seat 22 inches
Depth of Seat 18 inches
Code Word—PEROU

$255

3046—Sheraton shield backs of interesting design. The treatment of the seat and legs is most unusual.

$320

3046 SC
Height 38 inches
Across the Seat 21 inches
Depth of Seat 18 inches
Code Word—PERTY

3046 AC
Height 39 inches
Across the Seat 22 inches
Depth of Seat 19 inches
Code Word—PERUZ

Dining Room Furniture

1219 SC & AC—Ladderback Chippendale from the original owned by George Washington at Mount Vernon. Hand-carved.

$320

$275

1219 AC
Height 40 inches
Across the Seat 23 inches
Depth of Seat 19 inches
Code Word—ABATU

1219 SC
Height 39 inches
Across the Seat 21 inches
Depth of Seat 17 inches
Code Word—ABAST

1295 SC & AC—Sheraton's style usually followed along refined, simple lines, showing a strong tendency to the straight upright. An interesting feature of the models illustrated is the interlacing lines of the backs of the chairs.

$260

$295

1295 SC
Height 35 inches
Across the Seat 20 inches
Depth of Seat 17 inches
Code Word—ABKOA

1295 AC
Height 37 inches
Across the Seat 22 inches
Depth of Seat 18 inches
Code Word—ABKUF

SIMONDS

Dining Room Furniture

$295

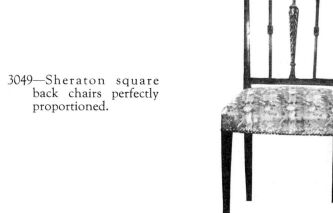

3049—Sheraton square back chairs perfectly proportioned.

$280

3049 AC
Height 35 inches
Across the Seat 21 inches
Depth of Seat 18 inches
Code Word—PESUA

3049 SC
Height 34 inches
Across the Seat 20 inches
Depth of Seat 17 inches
Code Word—PESSY

$345

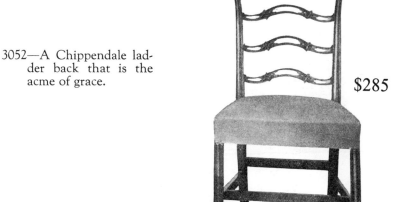

3052—A Chippendale ladder back that is the acme of grace.

$285

3052 AC
Height 39 inches
Across the Seat 23 inches
Depth of Seat 18 inches
Code Word—PETBI

3052 SC
Height 38 inches
Across the Seat 22 inches
Depth of Seat 17 inches
Code Word—PETAH

Dining Room Furniture

$270

2833 SC & AC—Developed from an antique model found in Baltimore.

$320

2833 SC
Height 37 inches
Across the Seat 21 inches
Depth of Seat 18 inches
Code Word—ORAGI

2833 AC
Height 38 inches
Across the Seat 22½ inches
Depth of Seat 19 inches
Code Word—ORAMO

$310

3034 SC & AC—Same as 2833 SC & AC except with slip seat instead of all upholstered seat. Also smaller than 2833 pattern.

$260

3034 AC
Height 34 inches
Across the Seat 22 inches
Depth of Seat 17 inches
Code Word—PEMTY

3034 SC
Height 34 inches
Across the Seat 18 inches
Depth of Seat 16 inches
Code Word—PEMOU

SIMONDS

Dining Room Furniture

1284 SC & AC—Adam chairs reproduced from an original in the collection of the Duke of Lancaster, in the South Kensington Museum in London.

$285

$320

1284 SC
Height 36 inches
Across the Seat 19 inches
Depth of Seat 17 inches
Code Word—ABAXY

1284 AC
Height 37 inches
Across the Seat 24 inches
Depth of Seat 19 inches
Code Word—ABAZA

1291 SC & AC—These chairs represent the full flower of XVIII Century design. Chinese Chippendale; hand-carved.

$300

$345

1291 SC
Height 37 inches
Across the Seat 22 inches
Depth of Seat 18 inches
Code Word—ABIEN

1291 AC
Height 38 inches
Across the Seat 23 inches
Depth of Seat 19 inches
Code Word—ABIFO

SIMONDS

Dining Room Furniture

$295

3031 SC and AC—Chippendale chairs from the Metropolitan Museum.

$355

3031 SC
Height 36 inches
Across the Seat 21 inches
Depth of Seat 17 inches
Code Word—PERCH

3031 AC
Height 37 inches
Across the Seat 22½ inches
Depth of Seat 18 inches
Code Word—PERDI

$295

3051—Chippendale pierced splat back chairs unusual in design, graceful proportion.

$355

3051 SC
Height 38 inches
Across the Seat 22 inches
Depth of Seat 17 inches
Code Word—PEREJ

3051 AC
Height 38 inches
Across the Seat 23 inches
Depth of Seat 17 inches
Code Word—PERIN

Dining Room Furniture

$355

3060 SC & AC—Chippendale design. Models from Harwich, England.

$310

3060 AC
Height 39 inches
Across the Seat 23 inches
Depth of Seat 18 inches
Code Word—PEVEN

3060 SC
Height 38 inches
Across the Seat 22 inches
Depth of Seat 17 inches
Code Word—PEVAJ

$310

3033 SC and AC—Hepplewhite model.

$375

3033 SC
Height 36 inches
Across the Seat 20 inches
Depth of Seat 16 inches
Code Word—PEMHI

3033 AC
Height 37 inches
Across the Seat 21½ inches
Depth of Seat 17 inches
Code Word—PEMIJ

Dining Room Furniture

$325

$265

3056—Hepplewhite splat back type slender and delicate in design from Middlesex County, England.

3056 AC
Height 39 inches
Across the Seat 22½ inches
Depth of Seat 18 inches
Code Word—PETHO

3056 SC
Height 38 inches
Across the Seat 21 inches
Depth of Seat 17 inches
Code Word—PETEL

$275

$355

3037—So called balloon back Colonial chairs. Originals made in Philadelphia after Hepplewhite designs.

3037 SC
Height 38 inches
Across the Seat 20 inches
Depth of Seat 16 inches
Code Word—PETIP

3037 AC
Height 40 inches
Across the Seat 22 inches
Depth of Seat 17 inches
Code Word—PETMU

Dining Room Furniture

$335

2909 SC & AC—"Bowknot" Chippendale dining-room chairs; hand-carved; ball-and-clawfeet.

$435

2909 SC
Height 37 inches
Across the Seat 22 inches
Depth of Seat 17 inches
Code Word—BUDEA

2909 AC
Height 40 inches
Across the Seat 23 inches
Depth of Seat 18 inches
Code Word—BUDIE

$335

1290 SC & AC—Magnificent examples of Chippendales dating from 1765; from originals in Victoria and Albert Museum, London. With acorn motif in splat, hand-carved.

$415

1290 SC
Height 37 inches
Across the Seat 21 inches
Depth of Seat 18 inches
Code Word—ABFAG

1290 AC
Height 38 inches
Across the Seat 22 inches
Depth of Seat 18 inches
Code Word—ABFYE

Dining Room Furniture

$415

3064 SC & AC—French Chippendale. Models from London.

$335

3064 AC
Height 38 inches
Across the Seat 24 inches
Depth of Seat 19 inches
Code Word—PETRY

3064 SC
Height 37 inches
Across the Seat 23 inches
Depth of Seat 18 inches
Code Word—PETOW

$320

3047 SC & AC—Queen Anne. Models from High Wycombe, England.

$375

3047 SC
Height 38 inches
Across the Seat 22 inches
Depth of Seat 15 inches
Code Word—PETTA

3047 AC
Height 39 inches
Across the Seat 22 inches
Depth of Seat 16 inches
Code Word—PETUB

SIMONDS

Dining Room Furniture

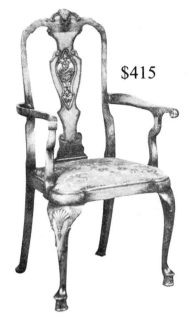

$415

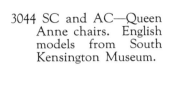

3044 SC and AC—Queen Anne chairs. English models from South Kensington Museum.

$335

3044 AC
Height 43 inches
Across the Seat 20½ inches
Depth of Seat 17 inches
Code Word—PERVA

3044 SC
Height 42 inches
Across the Seat 21 inches
Depth of Seat 15 inches
Code Word—PERYD

$425

3032 SC and AC—Chippendale chairs of beautiful proportion and detail.

$345

3032 AC
Height 42 inches
Across the Seat 23½ inches
Depth of Seat 18 inches
Code Word—PESAG

3032 SC
Height 40 inches
Across the Seat 25 inches
Depth of Seat 17 inches
Code Word—PESCI

Dining Room Furniture

$400

3028 AC
Height 48 inches
Across the Seat 24 inches
Depth of Seat 18 inches
Code Word—PASOR

$435

3028 T
Top 77 inches x 44 inches
Extends to 173 inches
Code Word—PASMO

$395

3028 SB
Height 38 inches
Top 72 inches x 22 inches
Code Word—PASIK

$300

3028 SC
Height 40 inches
Across the Seat 20 inches
Depth of Seat 18 inches
Code Word—PASSU

Dining Room Furniture

3028—A distinctive group of pieces individual in design and harmonious when assembled, having the sturdy charm of Jacobean craftsmanship.

$475

3028 Cabinet
Height 65 inches
Width 50 inches
Depth 19 inches
Code Word—PARZA

$390

3028½ AC
Height 42 inches
Across the Seat 23 inches
Depth of Seat 18 inches
Code Word—PEKED

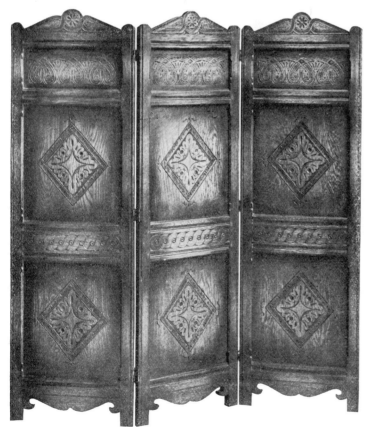

$300

$200

3028 ST
Height 30 inches
Top 44 inches x 24 inches
Code Word—PASAC

3028—Screen
Height 72 inches
Width 57 inches
Code Word—PASGI

Dining Room Furniture

$335

$425

3039—Early Georgian chairs with a wealth of exquisite detail. There is character in every line. Originals found in High Wycombe, England.

3039 SC
Height 37 inches
Across the Seat 24 inches
Depth of Seat 17 inches
Code Word—PESEK

3039 AC
Height 37 inches
Across the Seat 23½ inches
Depth of Seat 18 inches
Code Word—PESIO

$410

$325

3058—Chippendale fret back chairs sturdy and comfortable. The design is a combination of French and Chinese motives.

3058 AC
Height 39 inches
Across the Seat 23 inches
Depth of Seat 18 inches
Code Word—PESOV

3058 SC
Height 38 inches
Across the Seat 22 inches
Depth of Seat 17 inches
Code Word—PESNU

Occasional Chairs

$300

3366 AC—Quaint rendering of Queen Anne.

3367 AC—A Queen Anne model graceful in appearance, comfortable in seating quality.

$300

3366 AC
Height 36 inches
Across the Seat 23 inches
Depth of Seat 15 inches
Code Word—PEMIK

3367 AC
Height 35 inches
Across the Seat 23 inches
Depth of Seat 17 inches
Code Word—PEMMO

$290

3355 AC—Beautiful in every line is this Hepplewhite chair of many uses. The original model came from London.

3376 AC—Sheraton showing French influence. English model.

$290

3355 AC
Height 35 inches
Across the Seat 23 inches
Depth of Seat 17 inches
Code Word—PEMOR

3376 AC
Height 35 inches
Across the Seat 24 inches
Depth of Seat 17 inches
Code Word—PEMSU

Occasional Chairs

$420

2797 AC—Chippendale. Graceful in line, pleasing in proportion, this chair takes added interest by its unique upholstered arm and nail treatment.

2870 AC—Comfortable reading chair—Georgian detail.

$450

2797 AC
Height 36 inches
Across the Seat 27½ inches
Depth of Seat 20 inches
Code Word—OWAXE

2870 AC
Height 40 inches
Across the Seat 27 inches
Depth of Seat 21 inches
Code Word—OXAUC

$430

2785 AC—Chippendale again in fine proportion and rich detail.

2866 AC—Design of chair based on Jacobean motif.

$425

2866 AC
Height 46 inches
Across the Seat 26 inches
Depth of Seat 19 inches
Code Word—OUAJO

2785 AC
Height 42 inches
Across the Seat 27 inches
Depth of Seat 22 inches
Code Word—OXAPY

Occasional Chairs

$410

2882 AC & 2882½ AC—Queen Anne arm chairs different in line and of pleasing proportion.

$420

2882 AC
Height 36 inches
Across the Seat 23 inches
Depth of Seat 20 inches
Code Word—ONATS

2882½ AC
Height 36 inches
Across the Seat 23 inches
Depth of Seat 20 inches
Code Word—PATLO

$395

2845 AC—Interesting chair of small Coxwell type with Jacobean turnings and carved arms.

1642 AC—The rare dignity of this chair is borrowed from its early Jacobean prototype.

$410

2845 AC
Height 36 inches
Across the Seat 24 inches
Depth of Seat 20 inches
Code Word—OYAKU

1642 AC
Height 48 inches
Across the Seat 23½ inches
Depth of Seat 20 inches
Code Word—ALDIX

Occasional Chairs

$360

$350

4046 AC—Napoleon's ascendancy in the Empire period brought with it designs, such as this beautiful arm chair reminiscent of Rome and the Imperial Caesar.

4097 AC—Louis XVI from imported model.

4046 AC
Height 36 inches
Across the Seat 24 inches
Depth of Seat 19 inches
Code Word—FONHE

4097 AC
Height 35 inches
Across the Seat 22½ inches
Depth of Seat 18 inches
Code Word—OYAZI

$310

$295

4047 SC & AC—Directoire chairs of splendid construction and skillful design.

4047 AC
Height 35 inches
Across the Seat 22 inches
Depth of Seat 19 inches
Code Word—FONIF

4047 SC
Height 34 inches
Across the Seat 19 inches
Depth of Seat 16 inches
Code Word—ORAYA

Occasional Chairs

$400

2808 AC—Modern sag seat occasional chair suggesting Spanish influence.

2868 AC—English adaptation of Louis XV motifs.

$390

2808 AC
Height 36 inches
Across the Seat 23½ inches
Depth of Seat 19 inches
Code Word—OBACO

2868 AC
Height 34 inches
Across the Seat 23 inches
Depth of Seat 17 inches
Code Word—OXASA

$385

1641 AC—A popular type of modern design. Antique finish.

2637 AC—Flemish type with unusual back. Very comfortable sag seat.

$410

1641 AC
Height 35 inches
Across the Seat 24 inches
Depth of Seat 18 inches
Code Word—ALDET

2637 AC
Height 40 inches
Across the Seat 23½ inches
Depth of Seat 19 inches
Code Word—BODAR

Occasional Furniture

$375

2548 AC—Quaint Victorian occasional chair.

2724 AC—A Charles II Chair of occasional type, suitable for any room where no period is emphasized as well as those of strict early English.

$395

2548 AC
Height 31 inches
Across the Seat 22 inches
Depth of Seat 19 inches
Code Word—BINYT

2724 AC
Height 37 inches
Across the Seat 23 inches
Depth of Seat 19 inches
Code Word—BRYES

$400

3244 AC—Small Chippendale arm chair of correct line and proportion. Suitable for a great variety of rooms.

2636 AC—A charming example of Queen Anne furniture, covered in a handsome tapestry.

$410

3244 AC
Height 37 inches
Across the Seat 24 inches
Depth of Seat 20 inches
Code Word—PRYVU

2636 AC
Height 41 inches
Across the Seat 26½ inches
Depth of Seat 21 inches
Code Word—BOCTI

Occasional Chairs

$335

3286 AC—Hepplewhite tub chair. Copy of antique.

2933 AC—Louis XVI boudoir chair; antique finish.

$375

3286 AC
Height 33 inches
Across the Seat 22 inches
Depth of Seat 19 inches
Code Word—PEKFE

2933 AC
Height 36 inches
Across the Seat 23½ inches
Depth of Seat 20 inches
Code Word—BUINO

$320

3369 AC—English interpretation of modern design. Frame finished in black and gold.

3377 AC—Unique in proportion. Queen Anne design. Models from Sussex.

$325

3369 AC
Height 33 inches
Across the Seat 22 inches
Depth of Seat 17 inches
Code Word—PEJLI

3377 AC
Height 35 inches
Across the Seat 22 inches
Depth of Seat 14 inches
Code Word—PEJOL

Living Room Furniture

$475

2825 AC—Chippendale from an antique model.

2825½ AC—Same as above in loose cushion.

$475

2825 AC
Height 38 inches
Across the Seat 25 inches
Depth of Seat 20 inches
Code Word—PUPVO

2825½ AC
Height 34 inches
Across the Seat 25 inches
Depth of Seat 21 inches
Code Word—OLAXU

$410

4048 AC—This unique and charming piece was built around a swan motif that was found on an old Directoire arm chair.

2531 AC—Delightful occasional chair in the earlier Chippendale tradition. Shown here covered in fine wool tapestry with ornamental nails.

$400

4048 AC
Height 32 inches
Across the Seat 23 inches
Depth of Seat 18 inches
Code Word—FONLI

2531 AC
Height 35 inches
Across the Seat 24½ inches
Depth of Seat 19 inches
Code Word—BIKAS

Living Room Furniture

$450

1032 AC—Early Chippendale; the sweep back gives comfort to this ideal reading chair; moderately priced; adapted for the club or hotel as well as home.

1032½ AC—Comfortable chair for living room or bedroom.

$445

1032 AC
Height 38 inches
Across the Seat 26½ inches
Depth of Seat 22 inches
Code Word—AHATA

1032½ AC
Height 33 inches
Across the Seat 26½ inches
Depth of Seat 20 inches
Code Word—OXAWE

$465

2596 AC—A fine chair for any man's home. Beautiful in proportion and detail.

2603 AC—Same as 1032 but with wood arm.

$460

2596 AC
Height 40 inches
Across the Seat 27½ inches
Depth of Seat 19 inches
Code Word—BIYOU

2603 AC
Height 37 inches
Across the Seat 26½ inches
Depth of Seat 20 inches
Code Word—OYAFO

SIMONDS

Living Room Furniture

$485

1509 AC—"The Martha Washington" reproduced from an American Colonial original.

1509 AC
Height 40 inches
Across the Seat 28 inches
Depth of Seat 20 inches
Code Word—AKANY

$410

2644 AC—One of the finest expressions of the Colonial spirit we have ever produced. Grace and dignity in every line.

2644 AC
Height 37 inches
Across the Seat 24 inches
Depth of Seat 21 inches
Code Word—BOFAT

$415

2792 AC—One of Chippendale's most beauful creations showing slight French influence.

2792½ AC
Height 38 inches
Across the Seat 25½ inches
Depth 20 inches
Code Word—PATOS

$420

2792 AC
Height 38 inches
Across the Seat 25½ inches
Depth of Seat 20 inches
Code Word—OXAEM

Living Room Furniture

$255

1760 Ottoman & SC—
Companion pieces
with "The Elgin."

$345

1760 Ottoman
Height 17 inches
Top 25 inches x 20 inches
Code Word—AMOST

1760 SC
Height 34 inches
Across the Seat 21 inches
Depth of Seat 17 inches
Code Word—OZANY

$375

1760 AC—"The Elgin,"
a triumph of American furniture design;
solid walnut, extremely comfortable
and serviceable.

1760½ AC—Ideal Bridge
Chair. Higher and
shallower seat and
shorter arms than
1760 AC.

$395

1760 AC
Height 34 inches
Across the Seat 24 inches
Depth of Seat 21 inches
Code Word—AMOOP

1760½ AC
Height 35 inches
Across the Seat 24½ inches
Depth of Seat 20 inches
Code Word—OZAPA

Living Room Furniture

$490

2857 AC—Early Chippendale chair of graceful lines and enriched with carving.

$410

3200 AC—Refinement and comfort have been added to the quaintly simple outlines of this Queen Anne arm chair found in the Metropolitan Museum, New York.

2857 AC
Height 42 inches
Across the Seat 29 inches
Depth of Seat 21 inches
Code Word—OWATA

3200 AC
Height 36 inches
Across the Seat 24 inches
Depth of Seat 17 inches
Code Word—PUBIN

$435

3231 AC—Finely carved Queen Anne arm chair.

$410

3221 AC—Small Queen Anne chair of modern design. The use of restrained curves throughout makes this a most pleasing creation.

3231 AC
Height 39 inches
Across the Seat 26 inches
Depth of Seat 19 inches
Code Word—PUFVE

3221 AC
Height 32 inches
Across the Seat 25½ inches
Depth of Seat 23 inches
Code Word—PRYUT

SIMONDS

Living Room Furniture

3038 AC & SC—Queen Anne chairs from originals found in Ipswich, England.

$395

$340

3038 AC
Height 41 inches
Across the Seat 23½ inches
Depth of Seat 17 inches
Code Word—PENCE

3038 SC
Height 39 inches
Across the Seat 22 inches
Depth of Seat 17 inches
Code Word—PENAC

3331 & 3316 AC—The originals of these chairs from designs of William Kent is in the Victoria and Albert Museum in London.

$435

$430

3331 AC
Height 42 inches
Across the Seat 24 inches
Depth of Seat 19 inches
Code Word—PENEG

3316 AC
Height 41 inches
Across the Seat 24 inches
Depth of Seat 18 inches
Code Word—PENGI

Living Room Furniture

$410

2595 AC—Sturdy comfortable reading chair of modern lines.

2519½ AC—A simple high back chair of Charles II design. The comfort and beauty of this chair serves to make this chair highly desirable as an ascenting piece for colorful interiors of almost any period.

$455

2595 AC
Height 39 inches
Across the Seat 26½ inches
Depth of Seat 20 inches
Code Word—BIYIN

2519½ AC
Height 47 inches
Across the Seat 26 inches
Depth of Seat 19 inches
Code Word—OUADI

$445

1715 AC—The Italian interpretation of the style of Louis XV is shown in this chair of Lombardy origin.

2627 AC—Quaint occasional chair with the popular sag seat.

$435

1715 AC
Height 40 inches
Across the Seat 26 inches
Depth of Seat 25 inches
Code Word—AMEDU

2627 AC
Height 34 inches
Across the Seat 23½ inches
Depth of Seat 18 inches
Code Word—BOAKY

Living Room Furniture

$475

3292 AC—Solid walnut, extremely comfortable and serviceable.

3292 AC
Height 48 inches
Across the Seat 24½ inches
Depth of Seat 21 inches
Code Word—PEMBE

$495

2903 AC
Height 47 inches
Across the Seat 26 inches
Depth of Seat 22 inches
Code Word—PEMNU

$375

2608 AC—A modern chair for the living room developed along Sheraton lines.

2760 AC—Modern Queen Anne interpretation.

2608 AC
Height 39 inches
Across the Seat 27½ inches
Depth of Seat 22 inches
Code Word—BLIFY

$420

2760 AC
Height 38 inches
Across the Seat 26 inches
Depth of Seat 23 inches
Code Word—OYAIS

SIMONDS

Living Room Furniture

$345

2936 AC—Louis XV Bergere with squab cushion.

2965 AC—Late Louis XIV.

$355

2936 AC
Height 37 inches
Across the Seat 29 inches
Depth of Seat 22 inches
Code Word—BUIXY

2965 AC
Height 44 inches
Across the Seat 26½ inches
Depth of Seat 21 inches
Code Word—PUVON

$365

5046 AC—Directoire.

2962 AC—Louis XV Court chair; 33 inch seat; reminder of the days of voluminous Court costumes.

$375

5046 AC
Height 35 inches
Across the Seat 25½ inches
Depth of Seat 20 inches
Code Word—PUVUT

2962 AC
Height 40 inches
Across the Seat 33½ inches
Depth of Seat 24 inches
Code Word—BUMYD

English Models

$355

3375 SC & AC—Carolean period. Models from England.

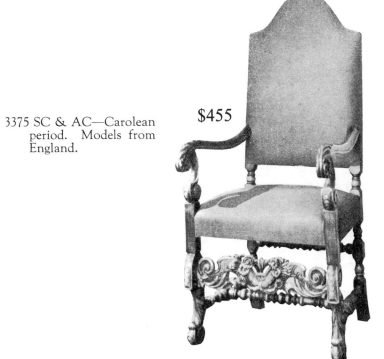

$455

3375 SC
Height 46 inches
Across the Seat 19 inches
Depth of Seat 15 inches
Code Word—PEMYA

3375 AC
Height 49 inches
Across the Seat 24 inches
Depth of Seat 17 inches
Code Word—PEMAD

$365

4049 SC & AC—Unusual in its fine proportions and refined detail. Louis XVI.

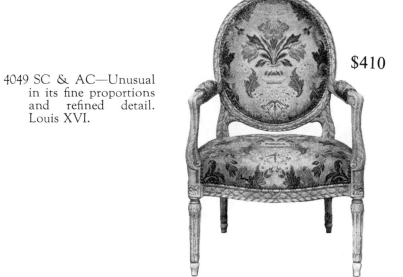

$410

4049 SC
Height 38 inches
Across the Seat 20 inches
Depth of Seat 18 inches
Code Word—PURAV

4049 AC
Height 39 inches
Across the Seat 26 inches
Depth of Seat 22 inches
Code Word—OBALY

Living Room Furniture

$275

4002 RS—A beautiful reproduction of the famous Savanarola Italian arm chair in walnut.

5020 AC—Richly carved Empire chair in Walnut and Gold.

$335

4002 RS
Height 34 inches
Across the Seat 17 inches
Depth of Seat 15 inches
Code Word—FOBEA

5020 AC
Height 28 inches
Across the Seat 24½ inches
Depth of Seat 18 inches
Code Word—OZAYI

5018 AC—Italian XVI Century.

2928½ AC—Louis XV taken from an antique French model.

$310

$310

5018 AC
Height 34 inches
Across the Seat 13½ inches
Depth of Seat 15 inches
Code Word—OCAES

2928½ AC
Height 35 inches
Across the Seat 23½ inches
Depth of Seat 19 inches
Code Word—OZAUE

Chairs and Table

$300

$310

3313—A modern English combination fireside chair which can be used as a tea table upon occasion.

3324—A low fireside chair. Charles II in design.

3313 Table Chair
Height 29 inches
Across the Seat 21 inches
Depth of Seat 16 inches
Table Height 21 inches
Table Top 21 inches x 21 inches
Code Word—PEXIU

3324 AC
Height 31 inches
Across the Seat 24 inches
Depth of Seat 20 inches
Code Word—PEXMY

$265

$295

3330—A low seated fireside chair that can be turned into a coffee table upon occasion.

3330 Table
Height 21 inches
Top 21 inches x 21 inches
Code Word—PEXOA

3330 AC
Height 28 inches
Across the Seat 21 inches
Depth of Seat 15 inches
Code Word—PEXOA

English Models

$295

$275

3338 AC—A quaint English chair perfect for card games.

3320 WC
Height 45 inches
Across the Seat 20 inches
Depth of Seat 14 inches
Code Word—PEXAL

3338 AC
Height 31 inches
Across the Seat 18½ inches
Depth of Seat 15 inches
Code Word—PEXEP

$490

3320 LS
Height 46 inches
Length overall 54 inches
Depth of Seat 14 inches
Code Word—PEXDO

3320—A small wood back chair and settee suitable for den or summer cottage.

Living Room Furniture

$400

3211 AC & Ott—Modern pieces built on lines of comfort and ornamented with brass nails.

$175

3211 AC
Height 36 inches
Across the Seat 25½ inches
Depth of Seat 21 inches
Code Word—PULU

3211 Ott
Height 18 inches
Top 24 inches x 22 inches
Code Word—PENYA

$310

1777 AC—A modern English easy chair of unusual lines. Especially adapted to the small room.

3247 AC—Different and pleasing is this small easy chair. Graceful cabriole legs with acanthus leaf carving. Back with attached pillow and loose cushion seat.

$300

1777 AC
Height 36 inches
Across the Seat 26 inches
Depth of Seat 21 inches
Code Word—AMITN

3247 AC
Height 33 inches
Across the Seat 24½ inches
Depth of Seat 18 inches
Code Word—PULZO

SIMONDS

Living Room Furniture

$185

1704 AC & Ottoman—
Models of the Coxwell
type. Louis XV motif.

$400

1618 or 1704 OTTOMAN
Height 17 inches
Top 26 x 26 inches
Code Word—PULTI

1704 AC
Height 38 inches
Across the Seat 26½ inches
Depth of Seat 25 inches
Code Word—AMBOG

$420

2806 AC—Coxwell chair
with design motifs
taken from a sofa
owned by the Duke of
Devonshire, England.

1795 AC—Modified
Louis XV Coxwell
chair with roll back
and loose cushion.

$410

2806 AC
Height 36 inches
Across the Seat 26 inches
Depth of Seat 23 inches
Code Word—OUAZE

1795 AC
Height 36 inches
Across the Seat 27 inches
Depth of Seat 22 inches
Code Word—AMVRY

Living Room Furniture

$390

1671 AC & Ottoman—
A Coxwell easy chair and Ottoman which are sweeping America in popular appreciation.

1651 AC & Ottoman—
Same as 1671 pattern except square back instead of roll top.

$210

1671 Ottoman
Top 26 inches x 22 inches
Height 18 inches
Code Word—OCAYL

1671 AC
Height 38 inches
Across the Seat 26 inches
Depth of Seat 26 inches
Code Word—ALGIA

$410

$395

2869 AC
Height 36 inches
Across the Seat 26 inches
Depth of Seat 24 inches
Code Word—OUAVA

2871 AC
Height 36 inches
Across the Seat 26½ inches
Depth of Seat 24 inches
Code Word—OUAYD

2869 AC—Coxwell type with legs, stretchers and arm supports in Charles II design.

2871 AC—Coxwell type. William & Mary design.

Living Room Furniture

$450

3288 AC—Queen Anne round back type.

2783 WC—Queen Anne with Clover Leaf back shaped for comfort.

$465

3288 AC
Height 40 inches
Across the Seat 25½ inches
Depth of Seat 21 inches
Code Word—PEPIM

2783 WC
Height 44 inches
Across the Seat 29½ inches
Depth of Seat 21 inches
Code Word—OWAHO

$500

3225 WC—Fine example of Colonial wing chair taken from original model in a Philadelphia collection.

2889 WC—Charles II wing chair with tufted back.

$500

3225 WC
Height 44 inches
Across the Seat 28 inches
Depth of Seat 21 inches
Code Word—PASVA

2889 WC
Height 37 inches
Across the Seat 27 inches
Depth of Seat 19 inches
Code Word—OMAIG

Living Room Furniture

$445

2668 WC—American Colonial chair showing Sheraton influence. Original in Metropolitan Museum, N. Y.

$445

2668½ WC
Height 42 inches
Across the Seat 30½ inches
Depth of Seat 20 inches
Code Word—PATAD

2668 WC
Height 42 inches
Across the Seat 30½ inches
Depth of Seat 20 inches
Code Word—BOFYR

$475

2768 WC—American wing chair.

3228 WC—Colonial and exact copy of an antique model.

$475

2768 WC
Height 39 inches
Across the Seat 27 inches
Depth of Seat 20 inches
Code Word—OVAYE

3228 WC
Height 44 inches
Across the Seat 29 inches
Depth of Seat 22 inches
Code Word—PUMNE

SIMONDS

Living Room Furniture

2939 AC—A comfortable Louis XV chair with loose down cushion.

$465

1618 WC—From imported French Louis XV model.

$460

2939 AC
Height 40 inches
Across the Seat 29 inches
Depth of Seat 24 inches
Code Word—BUJMO

1618 WC
Height 45 inches
Across the Seat 27 inches
Depth of Seat 22 inches
Code Word—ALBER

2855 WC—Queen Anne wing chair from an early example.

$475

3230 WC—Handsome Georgian wing chair with turned stretcher.

$475

2855 WC
Height 38 inches
Across the Seat 24½ inches
Depth of Seat 20 inches
Code Word—OWAEL

3230 WC
Height 46 inches
Across the Seat 30 inches
Depth of Seat 20 inches
Code Word—PRYJI

English Models

3319—A comfortable Wing chair without the apparent bulk of most English design; from Ipswich, England.

$410

$400

3363 WC—The Wing chair reduced to its smallest dimensions is fine for small rooms.

3319 WC
Height 41 inches
Across the Seat 21½ inches
Depth of Seat 17 inches
Code Word—PEWNY

3363 WC
Height 36 inches
Across the Seat 22 inches
Depth of Seat 17 inches
Code Word—PEWOZ

$410

3314—A delightful English Queen Anne Wing chair; original from Victoria Albert Museum.

3317—This simple Wing chair of English origin is a delightful piece for any man's summer home.

$420

3314 WC
Height 42 inches
Across the Seat 29 inches
Depth of Seat 21 inches
Code Word—PEWPA

3317 WC
Height 47 inches
Across the Seat 27 inches
Depth of Seat 19 inches
Code Word—PEWYI

Living Room Furniture

$440

1617½ AC—An Italian chair illustrating the southern influence on the style of the Grand Monarque.

1567 AC—Italian XVII Century.

$460

1617½ AC
Height 49 inches
Across the Seat 25 inches
Depth of Seat 21 inches
Code Word—ALBAM

1567 AC
Height 50 inches
Across the Seat 25 inches
Depth of Seat 21 inches
Code Word—AKEJY

$445

2956 AC—This rich Italian Renaissance piece with interlaced stretchers and carved crownpiece lends dignity to hall or library.

2955 AC—Late Italian Renaissance; heavy foliated carving; pierced finials, and paw feet.

$465

2956 A C
Height 55 inches
Across the Seat 26½ inches
Depth of Seat 19 inches
Code Word—BULYC

2955 AC
Height 54 inches
Across the Seat 26 inches
Depth of Seat 18 inches
Code Word—BULOT

Living Room Furniture

$465

1606 WC—A graceful reproduction of an old wing chair from Northern Italy. The period is late 17th century.

2771 WC—Unique Queen Anne wing chair and Ottoman.

$430

1606 WC
Height 43 inches
Across the Seat 26 inches
Depth of Seat 20 inches
Code Word—ALAXI

2771 WC
Height 41 inches
Across the Seat 26 inches
Depth of Seat 21 inches
Code Word—OWAIP

$455

$600

2606 AC
Height 47 inches
Across the Seat 25 inches
Depth of Seat 20 inches
Code Word—BLIAT

6013 LS
Height 35 inches
Length 52 inches
Depth of Seat 22 inches
Code Word—OLABY

2606 AC—Very dignified and massive library chair. The legs are particularly fine.

6013 LS—Modern Sofa built on lines of comfort and grace.

Living Room Furniture

$425

3306 WC—An American Colonial barrel chair copied from a Philadelphia antique.

3380 AC—Comfortable easy chair. Model from England.

$400

3306 WC
Height 41 inches
Across the Seat 27 inches
Depth of Seat 22 inches
Code Word—PEMVA

3380 AC
Height 35 inches
Across the Seat 28 inches
Depth of Seat 22 inches
Code Word—PEMAC

3268 AC—Formal chair of Umbrain type.

3334 LS—A love seat with wings, Queen Anne style. Quite different from the usual thing. Original purchased in the Marly-bone district of London.

$465

$550

3268 AC
Height 49 inches
Across the Seat 26 inches
Depth of Seat 19 inches
Code Word—PEMGI

3334 LS
Height 44 inches
Length overall 39 inches
Depth of Seat 18 inches
Code Word—PEMEG

Living Room Furniture

$410

1672 AC—A high type of modern American design. An easy chair that lives up to its name.

3360 AC & Sofa—Beautiful variant of the Martha Washington.

$425

1672 AC
Height 37 inches
Across the Seat 27 inches
Depth of Seat 21 inches
Code Word—ALGRI

3360 AC
Height 44 inches
Across the Seat 27 inches
Depth of Seat 20 inches
Code Word—PEHLI

$545-650

3360 Settee
Height 44 inches   Length 52 inches   Depth of Seat 20 inches
Code Word—PEHOL

Living Room Furniture

$395

2844 AC—Design suggested by an Italian antique chair.

$425

1741 AC
Height 35 inches
Across the Seat 25 inches
Depth of Seat 20 inches
Code Word—AMFIA

2844 AC
Height 42 inches
Across the Seat 25 inches
Depth of Seat 20 inches
Code Word—OZAIT

$320

$690

1741 SC
Height 34 inches
Across the Seat 20 inches
Depth of Seat 18 inches
Code Word—AMFEW

1741 Sofa
Height 35 inches
Length 54½ inches
Depth of Seat 22 inches
Code Word—AMGLE

1741 SC, AC & Settee—Of great distinction and historical interest are these pieces, reproduced from a magnificent Italian piece dating from the 17th century.

Living Room Furniture

$410

3362 AC—French Chippendale model from Oxford, England.

2639 AC—A Georgian type of small easy chair.

$395

3362 AC
Height 35 inches
Across the Seat 26 inches
Depth of Seat 19 inches
Code Word—PEPAE

2639 AC
Height 37 inches
Across the Seat 27 inches
Depth of Seat 22 inches
Code Word—BODOF

$475

$575

3213 AC
Height 36 inches
Across the Seat 25 inches
Depth 20 inches
Code Word—PUMDU

3213 LS
Height 36 inches
Width 45 inches
Depth 20 inches
Code Word—PUMEV

3213 Love Seat—Comfort and beauty are most happily wedded in this Chippendale love seat.

Living Room Furniture

$390

3379 CL
Height 31 inches
Length over all 53 inches
Length of Seat 46 inches
Across the Seat 25 inches
Code Word—PEJSO

3379 CL—Simple in design. Will fit in most bedrooms.

2634 Sofa—Simple Queen Anne kidney-shaped love seat with acanthus carving on legs.

$510

$515

2634 Sofa
Height 34 inches
Length 45 inches
Depth of Seat 20 inches
Code Word—BOCAP

3218 LS
Height 27 inches
Width 42 inches
Depth of Seat 21 inches
Code Word—PUDMU

3218 Love Seat—Low cozy fireside love seat original in design. The effective use of nail ornamentation makes this a most decorated piece.

Living Room Furniture

$410

3234 CL—Chaise Longue of graceful Louis XV form.

$400

3234 AC
Height 35 inches
Across the Seat 28½ inches
Depth of Seat 22 inches
Code Word—PUMIZ

3234 CL
Height 30 inches
Length 52 inches
Across the Seat 24½ inches
Code Word—PUPEY

3234 AC—Gracefully flowing lines of Louis XV adapted to the comfortable chair so much in demand today.

$690

$420

2853 Sofa
Height 48 inches
Length 57 inches
Depth of Seat 21 inches
Code Word—OJAZU

2853 Sofa & AC—Queen Anne reproductions.

2853 AC
Height 45 inches
Across the Seat 32 inches
Depth of Seat 21 inches
Code Word—OJAYT

SIMONDS

Living Room Furniture

$410

3344 WC—English model, small size.

3337 WC—Quite unusual is this armless wing chair. Has fine carving. Charles II period.

$440

3344 WC
Height 36 inches
Across the Seat 24½ inches
Depth of Seat 18 inches
Code Word—FEBKA

3337 WC
Height 45 inches
Across the Seat 23 inches
Depth of Seat 20 inches
Code Word—PEBOE

$495

$895

3318 WC
Height 49 inches   Across the Seat 26 inches
Depth of Seat 20 inches
Code Word—PEBTI

3318 Sofa
Height 48 inches   Length overall 51 inches
Depth of Seat 22 inches
Code Word—PEBZO

3318—These formal Carolean pieces are fitting frames for beautiful fabrics.

Living Room Furniture

$455

$710

3335 AC
Height 41 inches
Across the Seat 26 inches
Depth of Seat 18 inches
Code Word—PEVFO

3335 LS
Height 41 inches   Length over all 51 inches   Depth of Seat 19 inches
Code Word—PEVIS

3335 AC & LS Sofa—William and Mary design from English models.

$825

3335 Sofa
Height 41 inches   Length over all 84 inches   Depth of Seat 21 inches
Code Word—PEVKU

Living Room Furniture

$420

3326 WC—A small size Chippendale Wing chair. English model.

$400

3326 WC
Height 41 inches
Across the Seat 22 inches
Depth of Seat 18 inches
Code Word—PEDEW

3327 EC
Height 32 inches
Across the Seat 29 inches
Depth of Seat 25 inches
Code Word—PEDIA

3327 Sofa and Arm Chair—English club style, large, luxurious, and inviting to comfort.

$500

3327 Sofa
Height 32 inches
Length overall 75½ inches
Depth of Seat 26 inches
Code Word—PEDME

Living Room Furniture

$380

3371 WC—A modern English lounging chair, low and exceedingly comfortable.

$420

3371 WC
Height 33 inches
Across the Seat 22 inches
Depth of Seat 20 inches
Code Word—PEDXO

3328 AC
Height 31 inches
Across the Seat 29 inches
Depth of Seat 26 inches
Code Word—PEDYP

3328 Sofa and Arm Chair—English club style, large, luxurious, and inviting to comfort.

$520

3328 Sofa
Height 31 inches
Length overall 75½ inches
Depth of Seat 26 inches
Code Word—PEFAU

Living Room Furniture

$390

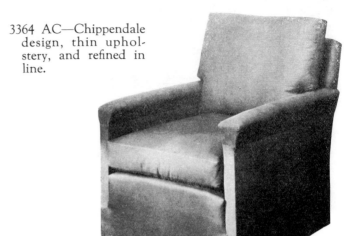

3364 AC—Chippendale design, thin upholstery, and refined in line.

$370

3364 AC
Height 39 inches
Across the Seat 28½ inches
Depth of Seat 20 inches
Code Word—PEDOG

3339 AC
Height 33 inches
Across the Seat 26 inches
Depth of Seat 23 inches
Code Word—PEDRI

3339 Sofa and Arm Chair—English all upholstered pieces of correct proportions for style and comfort.

$420

3339 Sofa
Height 35 inches
Length overall 75 inches
Depth of Seat 24 inches
Code Word—PEDUL

SIMONDS

Living Room Furniture

$320

3301 AC—Queen Anne round back arm chair, unusual in outline and proportion.

$380

3301 AC
Height 36 inches
Across the Seat 20 inches
Depth of Seat 20 inches
Code Word—PEFEY

3351 EC
Height 30 inches
Across the Seat 24 inches
Depth of Seat 22 inches
Code Word—PEFGA

3351 Sofa and EC—These pieces give the comfort of large English club furniture without their extreme bulk.

$480

3351 Sofa
Height 31 inches
Length overall 74½ inches
Depth of Seat 21 inches
Code Word—PEFIC

Living Room Furniture

3315 WC—Simple Queen Anne. Moderate in size and shape.

$450

$390

3378 AC
Height 39 inches
Across the Seat 29 inches
Depth of Seat 20 inches
Code Word—PEFOI

3315 WC
Height 42 inches
Across the Seat 25 inches
Depth of Seat 19 inches
Code Word—PEFKE

3378 Sofa & AC—Formal type of Queen Anne.

$700

3378 Sofa
Height 42 inches
Length 77 inches
Depth of Seat 23 inches
Code Word—PEFUN

Living Room Furniture

$775

2867 Sofa

Height 38 inches   Seat Rail 63 inches   Between Arms 4 feet 9 inches   Depth of Seat 23 inches
Code Word—ONAYX

2867 Sofa—Colonial Empire Sofa taken from an old model.

$390 $525

1670 AC & Sofa

Height 35 inches   Height 35 inches
Across the Seat 29 inches   Between Arms 6 ft. 6 inches
Depth of Seat 26 inches   Depth of Seat 28 inches
Code Word—ALGEW   Code Word—ALGAS

1670 AC & Sofa—Expressive of luxury and great comfort are this arm chair and sofa. The details are modified Italian, but in durability and in excellence of construction, the suite is essentially modern.

Living Room Furniture

$340

3219½ AC—American chair with Sheraton leg.

$380

3219½ AC
Height 34 inches
Width 25 inches
Depth of Seat 20 inches
Code Word—PUDTA

3201 AC
Height 34 inches
Width 25 inches
Depth of Seat 19 inches
Code Word—PUDRY

3201 Sofa and AC—Queen Anne group of most pleasing lines nicely tailored and comfortable.

$690

3201 Sofa
Height 37 inches
Length 72 inches
Depth of Seat 21 inches
Code Word—PUDOW

Living Room Furniture

3219 Sofa
Height 37 inches
Length 72 inches
Between Arms 60 inches
Depth of Seat 22 inches
Code Word—PUFAJ

$800

3219 & 3219½ Sofas—American Colonial sofa of Chippendale lines.

3219½ Sofa
Height 35 inches
Length 72 inches
Between Arms 60 inches
Depth of Seat 22 inches
Code Word—PUDUB

$795

Living Room Furniture

$650

$400

2878 LS
Height 34 inches
Length 38½ inches
Depth of Seat 18 inches
Code Word—PREDI

1702 EC
Height 36 inches
Across the Seat 28½ inches
Depth of Seat 28 inches
Code Word—AMACO

2878 LS—A small love seat of Charles II design.

$650

1702 Sofa
Height 38 inches
Length 91 inches
Depth of Seat 27 inches
Code Word—AMALY

1702 EC and Sofa—A Chippendale sofa and easy chair of unusual grace. The wings, combined with the great depth of seat are especially interesting. Made from a distinguished English model. Down cushions.

Living Room Furniture

2543 WC—Fireside chair of XVII Century ancestry, with William and Mary Stretchers. The wide "ears" or wings were originally designed as protection against draughts.

$490

$420

2543 WC
Height 45 inches
Across the Seat 30½ inches
Depth of Seat 21 inches
Code Word—BIMYS

1681 EC
Height 36 inches
Across the Seat 28 inches
Depth of Seat 24 inches
Code Word—ALHUM

1681 EC, Sofa and Pillows—Chippendale feeling; down-filled compartment cushions and spring edge; adapted for larger living rooms.

$580

1681 Sofa
Height 34 inches
Length 6 feet 6 inches
Depth of Seat 24 inches
Code Word—ALIUN

SIMONDS

Living Room Furniture

$400

1568½ AC
Height 35 inches
Across the Seat 27 inches
Depth of Seat 24 inches
Code Word—AKFAP

$650

1568½ Sofa
Height 35 inches
Length 6 feet 6 inches
Depth of Seat 24 inches
Code Word—AKFOE

1568½ AC & Sofa—"The Ambassador"; William and Mary influence. Adapted for club and library use.

$765

1568½ LS
Height 35 inches
Length 54 inches
Depth of Seat 24 inches
Code Word—AKFUJ

1568½ LS—A Love Seat developed from the "Ambassador" suite.

Living Room Furniture

$450

6005 AC—Modern Engglish Coxwell chair of great popularity and worth.

$475

6005 AC
Height 35 inches
Across the Seat 26 inches
Depth of Seat 26 inches
Code Word—HIBSU

1568 AC
Height 35 inches
Across the Seat 27 inches
Depth of Seat 24 inches
Code Word—AKELA

1568 Sofa & Arm Chair—These handsome and dignified pieces are historically correct interpretations of the style in Italy during the 18th Century.

$795

1568 Sofa
Height 35 inches
Length 6 feet 6 inches
Depth of Seat 24 inches
Code Word—AKEPE

Living Room Furniture

$495

2659 Group—Sofa and chairs designed for the room of moderate size. These pieces embody all the charm of their Charles II prototypes.

$530

2659 EC
Height 38 inches
Across the Seat 27½ inches
Depth of Seat 21 inches
Code Word—OCAMA

2659 WC
Height 45 inches
Across the Seat 27½ inches
Depth of Seat 21 inches
Code Word—OCARE

$895

2659 Sofa
Height 33 inches
Length 73 inches
Depth of Seat 23 inches
Code Word—BOISE

Living Room Furniture

$780

1744 LS
Height 37 inches
Length 44 inches
Depth of Seat 21 inches
Code Word—AMIJE

1744 LS—It is impossible to convey by illustrations the great charm of this Charles II love seat.

$720

2659 LS
Height 34 inches
Length 50 inches
Depth of Seat 22 inches
Code Word—OCAVI

Page Ninety-seven

Living Room Furniture

$420

2649½ AC
Height 37 inches
Width 25 inches
Depth 21 inches
Code Word—PATEH

$410

2826½ TC
Height 34 inches
Width 26 inches
Depth 21 inches
Code Word—PATIL

$895

2649½ Sofa
Height 36 inches
Length 72 inches
Depth of Seat 22 inches
Code Word—PATFI

2649½ Sofa & A C—Like 2649, but with cushion.

SIMONDS

Living Room Furniture

$400

2649 AC—Delicacy and grace of modeling and mastery of line are predominating features of this Sheraton Arm Chair.

$375

2649 AC
Height 37 inches
Across the Seat 25 inches
Depth of Seat 21 inches
Code Word—BOHJE

2826 TC
Height 34 inches
Across the Seat 26 inches
Depth of Seat 21 inches
Code Word—OCAIW

$895

2649 Sofa
Height 36 inches
Length 6 feet
Depth of Seat 22 inches
Code Word—BOHNI

2649 Sofa—A typical example of the charming, graceful lines of Sheraton.

SIMONDS

Living Room Furniture

$845

$410

5035 LS
Height 35 inches
Length 48½ inches
Depth of Seat 22 inches
Code Word—PUDEL

5035½ AC
Height 40 inches
Across the Seat 22 inches
Depth of Seat 19 inches
Code Word—PUDHO

5035 LS & 5035½ AC—Flemish influence.

$700

2626 Sofa
Height 34 inches
Length 71 inches
Depth of Seat 24 inches
Code Word—BOABO

2626 Sofa—Magnificent Spanish Sofa adapted from an old Cordova original

Living Room Furniture

$800

5035 Sofa
Height 34 inches
Length 72 inches
Depth of Seat 23 inches
Code Word—PUCUA

5035 Sofa, WC & EC—Richly carved and scrolled and comfortably upholstered pieces of Flemish influence.

$470

$530

5035 EC
Height 34 inches
Across the Seat 27 inches
Depth of Seat 22 inches
Code Word—PUDAH

5035 WC
Height 39 inches
Across the Seat 27½ inches
Depth of Seat 20 inches
Code Word—PUCYE

Living Room Furniture

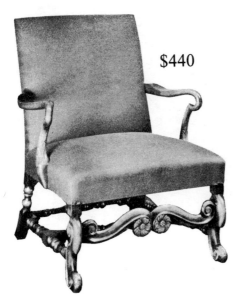

$440

3332 AC—Late Jacobean. Has comfort without heaviness.

$495

3332 AC
Height 37 inches
Across the Seat 28 inches
Depth of Seat 20 inches
Code Word—PEWAK

3382 AC
Height 38 inches
Across the Seat 27 inches
Depth of Seat 21 inches
Code Word—PEWEO

$800

3382 Sofa
Height 38 inches
Length overall 59 inches
Depth of Seat 23 inches
Code Word—PEWIT

3382 AC & Sofa—Chippendale, showing strong French influence.

SIMONDS

Living Room Furniture

$450

$400

3251 EC
Height 33 inches
Across the Seat 26 inches
Depth of Seat 28 inches
Code Word—PENMO

3282 AC
Height 36 inches
Across the Seat 29 inches
Depth of Seat 21 inches
Code Word—PENOR

3251 EC—Companion piece to 3251 Sofa.

3282 AC—Queen Anne deep seated, low and exceedingly comfortable.

$600

3251 Sofa
Height 33 inches
Width 72 inches
Depth of Seat 22 inches
Code Word—PUDIP

3251 Sofa—A clean cut compact sofa of late Queen Anne or early Chippendale design.

Living Room Furniture

$380

3255 AC & 3255½ AC—The Keystone. A favorite design for club and hotel.

$390

3255 AC
Height 37 inches
Across the Seat 25 inches
Depth of Seat 20 inches
Code Word—PATXA

3255½ AC
Height 37 inches
Across the Seat 25 inches
Depth of Seat 20 inches
Code Word—PAUKO

$600

1742 Sofa
Height 33 inches   Length 6 feet   Depth of Seat 21 inches
Code Word—AMIFA

1742 Sofa—Great refinement and the simplicity of elegance are expressed in this adaptation of late Louis XVI design.

Living Room Furniture

$400

3333 AC
Height 34 inches
Across the Seat 26½ inches
Depth of Seat 20 inches
Code Word—PEBAP

$430

3333 WC
Height 40 inches
Across the Seat 25 inches
Depth of Seat 20 inches
Code Word—PEBEU

3333 AC, WC & Sofa—Simple pleasing Jacobean lines embodying comfort as well as durability.

$750

3333 Sofa
Height 35 inches
Length 77 inches
Depth of Seat 21 inches
Code Word—PEBIY

Living Room Furniture

$450

3281 LS
Height 32 inches
Length overall 54 inches
Depth of Seat 22 inches
Code Word—PECHY

$400

3281 EC
Height 32 inches
Across the Seat 29 inches
Depth of Seat 22 inches
Code Word—PECEV

3281 Sofa—EC & LS—Finely constructed, down seats and backs. Lawson type.

$480

3281 Sofa
Height 32 inches
Length overall 76½ inches
Depth of Seat 23 inches
Code Word—PECIZ

SIMONDS

Living Room Furniture

$410

$390

1773 WC—No finer example of American craftsmanship in furniture has been produced within the last decade. In construction, comfort and appearance this piece closely approximates perfection.

2823 AC
Height 32 inches
Across the Seat 27½ inches
Depth of Seat 21 inches
Code Word—OBASE

1773 WC—Front Rail
Height 48 inches
Across the Seat 30 inches
Depth of Seat 20 inches
Code Word—AMSIM

2823 AC & Sofa—These unique and charming pieces were built around a swan motif that was found on an old Directoire arm chair.

$550

2823 Sofa
Height 32 inches
Length 6 feet
Depth of Seat 21 inches
Code Word—OBAUG

Living Room Furniture

$410

5036 AC—Early Empire.

$420

5036 AC
Height 34 inches
Across the Seat 23 inches
Depth of Seat 19 inches
Code Word—PUVPO

2500 WC
Height 36 inches
Across the Seat 28 inches
Depth of Seat 22 inches
Code Word—BIAHO

2500 WC & SOFA—Sheraton pieces of great distinction and grace, with classic wave-motif in the carving.

$690

2500 SOFA
Height 33 inches
Length 78 inches
Depth of Seat 24 inches
Code Word—BIAMU

Living Room Furniture

$410

3388 AC
Height 35 inches
Across the Seat 24 inches
Depth of Seat 21 inches
Code Word—PUMJA

$425

3248 AC
Height 35 inches
Across the Seat 26½ inches
Depth of Seat 21 inches
Code Word—PUNRI

3248 AC & 3249 Sofa—Sofa of Elizabethan inspiration. Linen fold panelling gives an air of distinction to this piece. Carving runs clear around frame.

$560

3249 Sofa
Height 35 inches
Width 72 inches
Depth of Seat 22 inches
Code Word—PUNUL

Living Room Furniture

$390

3384 AC—A delightful small tub chair reminiscent of the Empire period. The outside back and sides are in antique wood finish.

$400

3385 AC—An unusual treatment using late French motives. The wood sides and back are in antique wood finish.

3384 AC
Height 32 inches
Across the Seat 22 inches
Depth of Seat 20 inches
Code Word—PECNE

3385 AC
Height 35 inches
Across the Seat 23½ inches
Depth of Seat 21 inches
Code Word—PECOF

$380

3381 AC—An occasional small chair of simple lines finished around the back and sides in antique wood finish.

$410

3386 AC—A quaint chair of Queen Anne inspiration. This delightfully informal chair finds its place equally well in the bedroom, living room, or solarium. The wood sides and back are in antique finish.

3381 AC
Height 35 inches
Across the Seat 23 inches
Depth of Seat 21 inches
Code Word—PECJA

3386 AC
Height 34 inches
Across the Seat 22½ inches
Depth of Seat 21 inches
Code Word—PECSI

Solarium Suites

$375

$390

4081 AC—A new and different atmosphere may be brought into a room by using these odd pieces of French influence. Painted to harmonize with material if desired.

4081 SC
Height 33 inches
Across the Seat 19 inches
Depth of Seat 15 inches
Code Word—OSAIL

4081 AC
Height 32 inches
Across the Seat 26½ inches
Depth of Seat 21 inches
Code Word—FRUEJ

$380

$465

4080 AC
Height 36 inches
Across the Seat 24½ inches
Depth of Seat 22 inches
Code Word—FRUDI

4081 LS
Height 34 inches
Length 54 inches
Depth of Seat 21 inches
Code Word—OSARU

4080 AC—This modern French chair, painted to harmonize with the upholstery may be used as an odd chair in rooms that seek to avoid period style.

SIMONDS

Living Room Furniture

$390

3239 AC—Unique modern arm chair with characteristic emphasis of structural form.

3240 AC—Modern chair with curved back and reeded legs.

$370

3239 AC
Height 29 inches
Across the Seat 24 inches
Depth of Seat 20 inches
Code Word—PUKLA

3240 AC
Height 33 inches
Across the Seat 23½ inches
Depth of Seat 19 inches
Code Word—PUKOD

$410

3215 AC—Adam. Louis XVI influence.

3241 AC—A familiar form restyled in the modern manner.

$390

3215 AC
Height 38 inches
Across the Seat 23½ inches
Depth of Seat 18 inches
Code Word—PUBOU

3241 AC
Height 33 inches
Across the Seat 25½ inches
Depth of Seat 22 inches
Code Word—PULAP

Living Room Furniture

$380

3223 AC—Somewhat reminiscent of the Italian handling of the Classic mode is this graceful arm chair with its scrolled arms and rounded form. Back bordered with ornamental brass nails.

$360

3223½ AC
Height 32 inches
Across the Seat 21½ inches
Depth of Seat 20 inches
Code Word—PATRU

3223 AC
Height 32 inches
Across the Seat 21½ inches
Depth of Seat 20 inches
Code Word—PUPGA

$325

4057 AC—This Provincial arm chair adds a note of color to a somber corner. Suitable for living room or bedroom.

3358 WC—Colonial in design, sturdy in construction. Is comfortable and pleasing to the eye.

$345

4057 AC
Height 38 inches
Across the Seat 25 inches
Depth of Seat 21 inches
Code Word—FOPPO

3358 WC
Height 39 inches
Across the Seat 26 inches
Depth of Seat 19 inches
Code Word—PEMUW

Living Room Furniture

2849 AC
Height 35 inches
Across the Seat 24½ inches
Depth of Seat 19 inches
Code Word—OKATO

$380

2849 LS
Height 32 inches
Length 38 inches
Depth of Seat 19 inches
Code Word—OKAUP

$550

3342 WC
Height 44 inches
Across the Seat 23 inches
Depth of Seat 17 inches
Code Word—PENIK

$385

2849 Sofa
Height 32 inches
Length 54 inches
Depth of Seat 19 inches
Code Word—OKAYU

$600

2849 AC, LS & Sofa—Sheraton with all the charm of the master designers art.

3342 WC—An English Wing Chair of exceptionally fine proportions.

Bed Room Furniture

$370

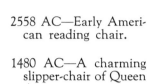

2558 AC—Early American reading chair.

1480 AC—A charming slipper-chair of Queen Anne inspiration.

$360

1480 AC
Height 30 inches
Across the Seat 24 inches
Depth of Seat 19 inches
Code Word—AJJEX

2558 AC
Height 40 inches
Across the Seat 23½ inches
Depth of Seat 19 inches
Code Word—BISUT

$310

2887 SC—Smart little bedroom chair.

3246 WC—Early American wing chair of quaint design.

$320

2887 SC
Height 28 inches
Across the Seat 24 inches
Depth of Seat 16 inches
Code Word—PREJO

3246 WC
Height 38 inches
Across the Seat 24 inches
Depth of Seat 19 inches
Code Word—PUCIO

Bed and Chest

$200

3502½ Bed
Height 41 inches
Height of Foot Board 19 inches
Side Rails 76 inches
Slats 39 inches
3 feet 3 inches Single Bed
Code Word—PEMZE

$300

3502 Chest
Height 50 inches
Width 42 inches
Depth 21 inches
Code Word—PEMAE

Vanities

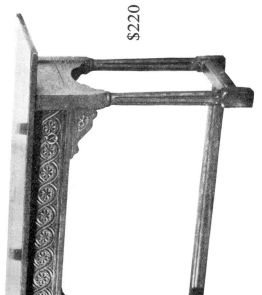

$220

3502 Vanity Table
Height 31 inches
Width 53 inches
Depth 20 inches
Code Word—PEMLO

3502 Hanging Mirror
Frame 36 inches by 23 inches
Glass 26 inches by 20 inches
Code Word—PEMUX

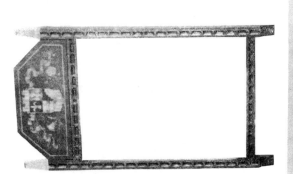

$45

3502 Bureau
Height 36 inches
Width 48 inches
Depth 21 inches
Code Word—PEMRU

$235

Vanities

3502 Suite—Combination of Elizabethan and Jacobean periods designed around a bed found in Uxbridge, England.

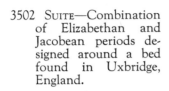

$210

3502—Toilet Mirror
Closed 25 x 18 inches
3 frames open 36 inches wide
Code Word —PEMOS

3502—Vanity Table
Height 31 inches
Width 53 inches
Depth 20 inches
Code Word—PEMLO

$90

$175

3502 Stand
Height 31 inches
Width 17 inches
Depth 14 inches
Code Word—PEMVY

SIMONDS

Vanities

$200

3116 DT
Height 29 inches
Width 37 inches
Mirror 13 x 12 inches
Code Word—PUTAX

3116 DT—Louis XVI designed for use in a room corner.

$290

3175 DT
Height 30 inches   Top 29 inches x 16 inches
Code Word—PECUK

3175 DT—Graceful vanity table. Venetian decoration.

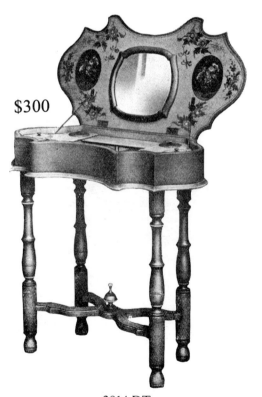

$300

$80

Top View 2814 DT

2814 DT—A gaily painted Vanity Table compact in size for boudoir or dressing room.

2814 DT
Height 30 inches
Top 30 inches x 15 inches
Code Word—OHAIB

Living Room Furniture

2828 LS
Height 36 inches
Length 52 inches
Depth of Seat 19 inches
Code Word—OBAER

$550

2828 LS—This love seat developed around the base of an old cabinet is of unusual interest.

2528 Sofa
Height 34 inches
Length 70 inches
Depth of Seat 25 inches
Code Word—BIGYL

$750

2528 Sofa—Exquisite interpretation of late Colonial type; solid mahogany with delicate crotch facings. These pieces are noteworthy for the distinction of their outline.

Living Room Furniture

3300 Sofa

Height 33 inches   Length 72 inches   Depth of Seat 22 inches
Code Word—PEVOY

3300—A graceful Chippendale sofa showing French motives.

3373 Sofa

Height 38 inches   Length 78 inches   Depth of Seat 26 inches
Code Word—PEVRA

3373 Sofa—The model for this beautiful French Hepplewhite sofa was found in High Wycombe, England.

Living Room Furniture

$650

3242 Sofa
Height 33 inches
Length overall 66 inches
Depth of Seat 23 inches
Code Word—PUCNU

3242 Sofa—Louis XV, very graceful and finely tailored.

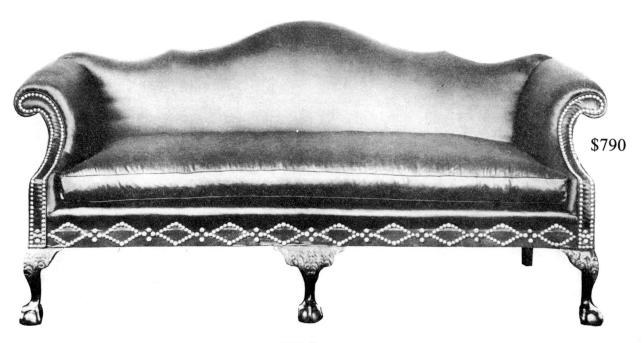

$790

3298 Sofa
Height 35 inches
Length overall 72 inches
Depth of Seat 22 inches
Code Word—PECAR

3298 Sofa—A comfortable Georgian sofa showing an interesting nail treatment.

Living Room Furniture

$645

3285 Sofa
Height 33 inches
Length 76 inches
Depth of Seat 23 inches
Code Word—PEZKY

3285 Sofa—Sheraton sofa. Well proportioned. Good lines and comfortable.

$640

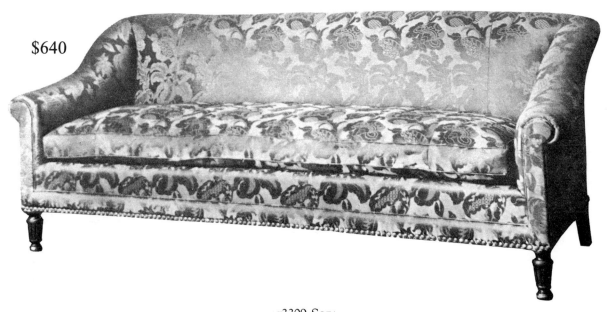

3309 Sofa
Height 31 inches
Length 76 inches
Depth of Seat 23 inches
Code Word—PEZMA

3309 Sofa—A deep comfortable sofa of sturdy English feeling with legs in Louis XVI style.

SIMONDS

Miscellaneous Chairs

$175

3207 SAC—A dainty, low, comfortable, revolving bedroom chair of Louis XVI.

2881 SC—Sheraton desk or dressing table chair.

$145

3207 SAC
Height 28 inches
Across the Seat 22 inches
Depth of Seat 17 inches
(Revolving)
Code Word—PUFEN

2881 SC
Height 32 inches
Across the Seat 17 inches
Depth of Seat 17 inches
(Revolving)
Code Word—PRUID

$135

2880 SC—Swivel chair in the style of Louis XVI. This chair will group with many fine bedroom suites.

2778 SAC—For the fine business or professional office.

$185

2880 SC
Height 34 inches
Across the Seat 18 inches
Depth of Seat 18 inches
(Revolving)
Code Word—OMAUS

2778 SAC
Height 37 inches
Across the Seat 22½ inches
Depth of Seat 20 inches
(Revolving)
Code Word—OLASO

Office Furniture

3296—Chippendale design.

3297½—Comfortable Office chair of conservative design.

3296 AC
Height 39 inches
Across the Seat 26 inches
Depth of Seat 20 inches
Code Word—PEHSO

3297½ Sw AC
Height 37 inches
Across the Seat 24 inches
Depth of Seat 18 inches
Code Word—PEHUR

3308 AC—Copied from signers chair in Independence Hall, Philadelphia.

3308½ Sw. AC—Office chair based on signers chair as shown under No. 3308.

3308 AC
Height 37 inches
Across the Seat 24 inches
Depth of Seat 18 inches
Code Word—PEICY

3308½ Sw AC
Height 37 inches
Across the Seat 24 inches
Depth of Seat 18 inches
Code Word—PEIMI

Office Furniture

$300

1760 LS
Height 34 inches
Length 42 inches
Depth of Seat 21 inches
Code Word—PAVYD

$265

3216 Sw AC
Height 22 inches
Across the Seat 22 inches
Depth of Seat 22 inches
Code Word—PEMOV

$245

3216 Sw SC
Height 22 inches
Across the Seat 22 inches
Depth of Seat 22 inches
Code Word—PEMSY

Office Furniture

$275

3258 AC
Height 35 inches
Across the Seat 25 inches
Depth of Seat 20 inches
Code Word—PAVOU

$275

3258½ Swivel AC
Height 34 inches
Across the Seat 23½ inches
Depth of Seat 19 inches
Code Word—PAVTY

$325

1760½ Swivel AC
Height 34 inches
Across the Seat 22 inches
Depth of Seat 18 inches
Code Word—PAVUZ

$220

1760 Swivel SC
Height 34 inches
Across the Seat 20 inches
Depth of Seat 17 inches
Code Word—PEMUZ

Office Furniture

$265

3287½ Swivel Arm Chair—Early English.

3290½ Arm Chair—Modern.

$265

3287½ Sw. AC
Height 34 inches
Across the Seat 22½ inches
Depth of Seat 20 inches
Code Word—PEITO

3290½ AC
Height 31 inches
Across the Seat 24 inches
Depth of Seat 19 inches
Code Word—PEJAW

$850

3157 Early English
Height 29 inches
Top 62 inches x 32 inches
Code Word—PEJBY

3157 Early English—Elizabethan design

Page One Hundred Twenty-eight

Occasional Pieces

$135

3329 S—A small Jacobean stool for the fireside corner.

$125

3329 Stool
Height 17 inches
Top 10 inches x 10 inches
Code Word—PEMAF

3345 Stool
Height 23 inches
Top 15 inches
Code Word—PEMOT

$210

3345—An ideal combination for bridge. Chairs are accessible and comfortable. Table is ample in size and has a leather playing surface.

$200

3345 AC
Height 31 inches
Across the Seat 19 inches
Depth of Seat 15 inches
Code Word—PEMWA

3345 Card Table
Height 27 inches
Top 33 inches x 30 inches
Code Word—PEMKO

Stools

$120

4041 Stool
Height 7 inches
Top 28 inches x 9 inches
Code Word—FONAX

$120

4039 Stool
Height 7 inches
Top 28 inches x 12 inches
Code Word—FOMIE

$100

3131 S—Elizabethan.

3131 Stool
Height 18 inches
Top 21 inches x 13 inches
Code Word—PUMSI

$95

4092 Stool
Height 8 inches
Top 20 inches x 9 inches
Code Word—FRUZE

Benches

$335

1741 Bench
Height 19 inches
Top 40 inches x 15 inches
Code Word—OTAKO

1741 Bench—Of great distinction and historical interest, reproduced from a magnificent Italian piece dating from the 17th century.

$320

3383 Bench
Height 17 inches
Top 39 inches x 17 inches
Code Word—PEYAM

3383 Bench—Early Queen Anne with an interesting leg treatment. The rails are veneered with walnut crossbanding.

Benches

$100

3272 Bench
Height 19 inches
Top 23 inches x 12 inches
Code Word—PEYCO

$155

1497 Bench
Height 18 inches
Length 56 inches
Width 14 inches
Code Word—AJUBE

3272 Bench—Sheraton.

$220

4042 Bench
Height 20 inches
Top 42 inches x 16 inches
Code Word—FONBY

$210

1603 Bench
Height 18 inches
Length 46 inches
Width 16 inches
Code Word—ALAOA

4042 Bench—Marie Antoinette piano bench of great beauty and refinement of detail.

1603 Bench—Italian interpretation of Louis XVI

Benches

$225

2690 Bench
Height 19 inches
Top 72 inches x 15 inches
Code Word—BOHAV

2690 Bench—A Spanish 17th Century Bench adapted from one in the Archaeological Museum in Madrid.

$230

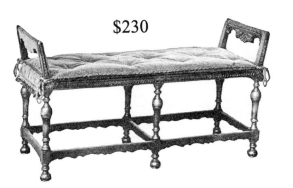

$200

1605 Bench
Height of Seat 20 inches
Width 18 inches
Length 48 inches
Code Word—ALATE

1581 Bench
Height 19 inches
Width 17 inches
Length 36 inches
Code Word—AKIWO

1581 Bench—This graceful bench of Adam design is especially appropriate for the boudoir.

Benches

$245

3283 BENCH
Height 36 inches   Length 60 inches   Depth of Seat 15 inches
Code Word—PEXTE

3283 BENCH—Spanish design.

$475

3275 SETTEE
Height 53 inches   Length overall 48 inches   Depth of Seat 18 inches
Code Word—PEXUF

3275 SETTEE—Elizabethan design.   Seat is hinged.

Table and Bench

3172 Bench
Height 18 inches
Top 60 inches x 15 inches
Code Word—PEYER

$220

3172—Quaint early American pine bench with Dutch type foot.

$400

3171 Table
Height 30 inches
Top 78 inches x 30 inches
Code Word—PEYHU

3171 Table—Pine table typical early American trestle table.

Tables and Stands

$145

3173 Table—Early colonial candle stand.

3193 Table—A reproduction of a Colonial stand found in Shedd's Corners, New York.

$165

3173 Table
Height 26 inches
Top 20 inches
Code Word—PEHDA

3193 Table
Height 26 inches
Top 18 inches
Code Word—PEHEB

$475

$140

2555 Gate Leg Table
Height 29 inches
Width 48 inches
Drawer 18 inches
Open 48 inches
Code Word—PEHIF

3522 CT
Height 16 inches
Top 25 inches
Code Word—PEMAG

2555 Table—Double gate leg Colonial table.

3522 Coffee Table—An interesting coffee table of the Queen Anne reign decorated with a Chinese raised lacquer painting in the typical manner of that age.

Cupboard

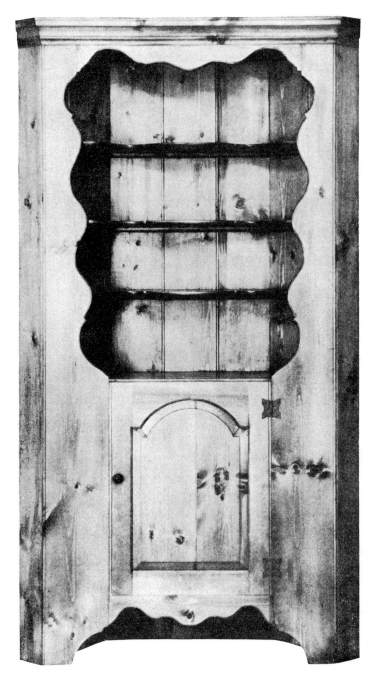

$900

3170 Cupboard
Height 87 inches
Width 40 inches
Depth 15 inches
Code Word—PEYLY

3170 Cupboard—Pine cupboard early 18th Century adapted from one in the Nutting collection.

Dresser

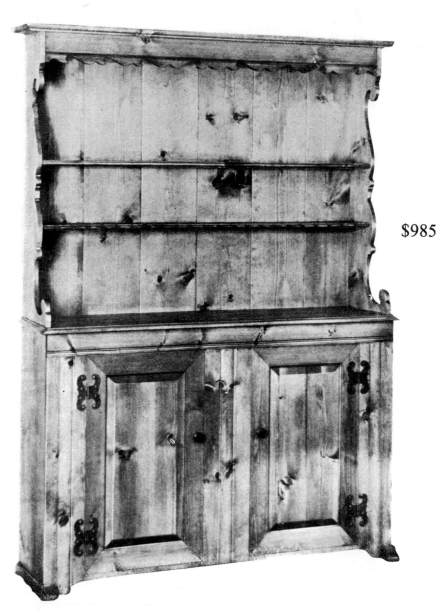

$985

3169 Dresser
Height 75 inches
Width 54 inches
Depth 15 inches
Code Word—PEYIV

3169 Dresser—Early 18th Century taken from a Pine dresser in a Boston collection.

Stands and Chests

$300

3112 S
Height 18 inches
Top 18 inches x 12 inches
Code Word—PUTBY

3112 Stool—The scrolled rail and incised carving in the turnings give this piece unusual interest.

3104 Stand—Jacobean stand with incised carved rail.

$290

3104 S
Height 20 inches
Top 20 inches x 11 inches
Code Word—PUTEB

$285

3110 S
Height 14 inches
Top 17 inches x 11 inches
Code Word—PUREZ

3110 Stool—Has a hinge lid. Jacobean.

$295

3145 C
Height 20 inches
Top 40 inches x 17 inches
Code Word—PAXRY

3145 Chest and 3164 Chest—Quaint Early English with linen fold sides.

These chests are made so that they can be used for piano rolls and phonograph albums.

$275

3164 C
Depth Inside 13 inches
Height 18 inches
Top 16 inches x 22 inches
Code Word—PEMEK

Tables and Stands

$345

2854 SS—This stand with its splayed legs, interesting "point twist" and carving is of Spanish inspiration but suitable to English and Italian interiors. Made in Oak, also Walnut.

2509 SS—Charles II decanter table with hand-carved details.

$290

2854 SS
Height 18 inches
Top 16 inches x 16 inches
Code Word—OFAJA

2509 SS
Height 23 inches
Top 20 inches
Code Word—BIDEO

$320

2692 ET—The burl walnut drawer front and top combine with the twist legs to give this William and Mary end table unusual charm and beauty.

2812 SS—Louis XVI stand of wide usefulness.

$290

2692 ET
Height 24 inches
Top 24 inches x 13 inches
Code Word—BOJDA

2812 SS
Height 21 inches
Top 15 inches
Code Word—OMAWU

Tables and Stands

$230

2632 Book Stand—Queen Anne in design. Unusually adaptable. Hand decorated.

2602 SS—Jacobean Smoking Stand. May be had with glass top.

$210

2632 BS
Height 24 inches
Top 22 inches x 20 inches
Code Word—BOBLA

2602 SS
Height 20 inches
Top 13 inches in diameter
Code Word—BLANY

2580 DLT—Drop leaves on three sides making a triangle top when down.

$210

$200

4081 Coffee Table
Height 19 inches
Top 25 inches x 17 inches
Code Word—OSAUX

2580 DLT
Height 28 inches
Top 30 inches open
Top 23 inches closed
In Maple with Decorations
Code Word—BIWAD

SIMONDS

Stands and Coffee Tables

$280

3120 T
Height 18 inches
Top 18 inches x 10 inches
Extended 24 inches
Code Word—PUTDA

3120 Table—Quaint and pleasing is this sturdy little drop leaf table. Jacobean.

3130 S—Elizabethan.

$220

3130 Stool
Height 20 inches
Top 24 inches x 17 inches
Code Word—PAXEL

$225

3101 Table
Height 21 inches
Top 26 inches x 17 inches
Code Word—PUSEA

3101 Table—Louis XVI.

3103 Table—Tripod with arched rails. Jacobean

$210

3103 Table
Height 22 inches
Top 18 inches
Code Word—PUTHE

Stands

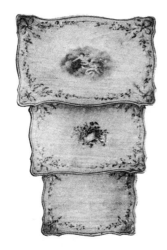

3524 TABLES—This delightful nest of tables is Sheraton in his most happy mood. Decorated and glazed in color.

3515 TABLE—Decorated on satinwood. English model.

$445

$240

3524 TABLES
Height 25 inches
Top 21 inches x 14 inches
Top 18 inches x 13 inches
Top 14 inches x 11 inches
Code Word—PEGUO

3515 TABLE
Height 26 inches
Top 28 inches x 18 inches
Code Word—PEGZU

3521 & 3525—Decorated coffee tables, Queen Anne design.

$320

$295

3521 TABLE
Height 20 inches
Top 19 inches x 16 inches
Code Word—PEHAX

3525 TABLE
Height 16 inches
Top 21 inches x 17 inches
Code Word—PEHBY

Tables and Stands

3514 Stand
Height 29 inches
Top 21 inches x 14 inches
Code Word—PEMIO

$240

2703 Coffee Table
Height 18 inches
Top 26 inches x 17 inches
Code Word—BROYC

$210

2703 CT—In keeping with the taste of the time, Sheraton made many an adaptation from the court of Marie Antoinette. The refined, graceful lines and hand painted top of this coffee table bespeak the inspiring influence of Louis XVI period upon the master designer.

3108 Coffee Table
Height 19 inches
Top 24 inches diameter
Code Word—PUFZI

$265

2701 CT
Height 20 inches
Top 32 inches x 20 inches
Code Word—BROOT

$210

3108 Table—Pompeian painted, decorated and ornamented with brass lion head mounts.

2701 Coffee Table—A unique Coffee Table with quaint Georgian legs and charming painted decorations.

Stands and Cabinets

$300

3529 Rev. BS
Height 26 inches
Diameter 18 inches
Code Word—PEGFA

3529 Stand—Queen Anne revolving with stationary top.

3195—A Chinese Chippendale urn stand delicate in detail.

$290

3195 Stand
Height 23 inches
Top 12 inches x 12 inches
Code Word—PEGID

3526 Coffee Table—A delightful painted table decorated with a floreated design. Interesting features of this table are the Flemish legs and the delicate pie crust top.

$275

3526 Coffee Table
Height 18 inches
Top 23 inches
Code Word—PEMCI

$375

3158 BC—Small revolving book case with handy locked cupboard.

3158 Rev. BC
Height 30 inches
Top 21 inches
Code Word—PEGNI

Tables

$275

3129 Table
Height 29 inches
Length 30 inches
Width 22 inches
Code Word—PUSTO

$245

3100 Table
Height 28 inches
Top 33 inches in diameter
Code Word—PUSOK

3129 Table—Charles II table with fancifully carved rail and turned legs.

3100 Table—Colonial table with Chippendale ball and claw feet. Top of selected mahogany.

$245

2786 Coffee Table
Height 20 inches
Top 32 inches x 20 inches
Code Word—OFAEV

$210

2840 Coffee Table
Height 20 inches
Top 32 inches x 21 inches
Code Word—OFAHY

2786 CT—Chippendale. With Japanese ash burl top.

2840 CT—Very unusual is this coffee table of French design with its interlacing and carved underframe.

Tables

$220

3507 Table—A simple English table. Staple on account of its three point contact with the floor.

3198 Stand—A convenient low stand for magazines or books.

$190

3507 Table
Height 23 inches
Top 27 inches
Code Word—PEMUY

3198 Stand
Height 17 inches
Top 19 x 19 inches
Code Word—PEMPU

$300

3520 Table
Height 27 inches
Top 31 inches
Code Word—PEMIN

3520 Table
Height 27 inches
Top 31 inches
Code Word—PEMIN

3520 Table—A combination game and utility table Chippendale design. Top decorated one side, cloth playing surface on reverse.

SIMONDS

Tables and Stands

2556 TTT—Early American "dish-top" tip table in maple with characteristic coaching scene. The decorations are applied by our own artists following historical models.

3513 STAND—Georgian.

$295

$275

2556 TTT
Height 25 inches
Top 24 inches in diameter
Code Word—BISON

3513 STAND
Height 30 inches
Top 12 inches
Shelf 17 inches
Code Word—PEGEZ

3165 & 3166—Authentic pie crust tripod tables with the edge of top cut from solid wood.

$375

$410

3165 TT
Height 30 inches
Top 30 inches
Code Word—PEGAV

3166 TT
Height 28 inches
Top 23 inches
Code Word—PEGDY

Tables and Cabinets

$235

1684 ET—A Jacobean end table in oak. This interesting piece harmonizes well with Italian furniture.

3192 Table—Hexagonal table of Elizabethan design.

$265

1684 ET
Height 25 inches
Top 31 inches x 14 inches
Code Word—ALJID

3192 Table
Height 29 inches
Top 31 inches
Code Word—PEFVO

$275

$380

2729 T
Height 27 inches
Top 33 inches x 30 inches
Folded 33 x 16 inches
Code Word—BRYMA

2628 Cabinet
Height 40 inches
Width 30 inches
Depth 13 inches
Code Word—BOAMA

2729 Table—The solidity of design and structure of this Jacobean gate-leg table in oak, makes it a piece of great distinction.

2628 Cabinet—An interesting piece of Jacobean design in oak. It fits in excellent harmony with Italian surroundings.

Tables

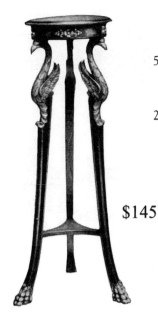

5021—Empire from imported model. Walnut and gold.

2824—Duncan Phyfe table beautifully carved and inlaid. Mahogany and satinwood.

$145

$290

5021 Pedestal
Height 42 inches
Top 13 inches
Code Word—OFAUK

2824 T
Height 28 inches
Top 30 inches
Code Word—OFAYO

3512 Table—A Sheraton Drum Table quite different in design and practical in size.

5020—Empire. Gilded ornaments.

$300

$280

3512 Table
Height 29 inches
Top 33 inches
Code Word—PEME

5020 RT
Height 31 inches
Top 26 inches
Code Word—OGAEW

Tables

$290

3113 Table—Jacobean table with turned legs and stopped flutes. A band of carving around the apron adds interest to this piece. Has one drawer in end.

$310

1497 SS
Height 23 inches
Length 16 inches
Width 12 inches
Code Word—AJUPT

3118 T
Height 25 inches
Length 22 inches
Width 16 inches
Code Word—PURO

$400

1497½ LT
Height 30 inches
Top 72 inches x 28 inches
Code Word—OHABU

1497 SS & 1497½ LT—Simple Jacobean design of great interest for clubs and interiors of simple design.

Tables

$380

1746¼ LT
Height 30 inches
Top 66 inches x 26 inches
Code Word—AMKEA

1746½ LT

1746½ LT
Height 30 inches
Top 72 inches x 30 inches
Code Word—AMKYU

1746½ & 1746¼ LIBRARY TABLES—An Italian table of dignified character. Aside from living room use, it is suitable for dining rooms.

$320

$275

2625 OT
Height 26 inches
Top 30 inches
Code Word—BLYSA

1746 OT
Height 30 inches
Top 32 inches
Code Word—AMIZU

2625 TABLE—Octagonal table developed from our 2509 Stand shown on page 140. Beautifully figured top. Charles II inspiration.

1746 TABLE—An Italian 16th century octagonal table, inspired by an old pedestal table in the Bardini collection in Florence.

Tables

$255

2817 T
Height 27 inches
Top 30 inches
Code Word—OGACU

2817 Table—William and Mary design, rich in inlay and decoration.

3119 Table—Quaint Jacobean table with three legs supporting eight sided top.

$265

3119 T
Height 25 inches
Top 24 inches in diameter
Code Word—PURID

$295

2848 T
Height 28 inches
Top 30 inches
Code Word—OGAOG

2848 Table—Richly carved table of Italian design. Oak.

2847 Table—Chas. II table in oak richly carved.

$275

2847 T
Height 28 inches
Top 28 inches
Code Word—OGARI

Tables

3134 Table
Height 29 inches
37 inches from corner to corner
32½ inches side to side
Code Word—PAXBI

$245

3134 & 3135 Tables—Simple Jacobean designs

$265

3135 Table
Height 29 inches
Top 42 inches x 22 inches
Code Word—PAXTA

Tables

$190

3121 Table
Height 23 inches
Top 21 inches in diam.
Code Word—PUJAM

3121 Table—Modern table in maple with top in leather of modern design.

3123 Table—Modern table of decorated maple. Top shelf is of heavy glass through which is seen modern decorations painted on wood shelf below.

$170

3123 Table
Glass Top
Height 23 inches
Center Shelf 28 in. diam.
Code Word—PUJOC

$265

3109 Table
Height 30 inches
Top 36 inches in diam.
Code Word—PUSYU

$300

3113 Table
Height 30 inches
Top 44 inches x 44 inches
Code Word—PARAB

3109 Table—A reproduction of an antique Duncan Phyfe reading table with revolving top.

3113 Table—Jacobean octagonal table with twisted turnings and carved ornament.

Tables

3141 Table
Height 30 inches
Top 54 inches x 42 inches
Folded 42 x 18 inches
Code Word—PEYNA

$500 EACH

3141—Early English twist gate leg table of oak. Good dining size.

3141 Table
Height 30 inches
Top 54 inches x 42 inches
Folded 42 x 18 inches
Code Word—PEYNA

Tables

$425

3136 Table
Top 44 inches in diameter
Height 29 inches
Code Word—PURZU

3136 OT—Made of mahogany. A simple Colonial table of Duncan Phyfe inspiration of ample size and good proportion finished with brass sockets of the period.

$450

1332 DLT
Height 31 inches
Top 60 inches x 32 inches
Folded 32 inches x 32 inches
Code Word—AFDOX

1332 DLT—A colonial drop leaf table—one of the finest examples of Duncan Phyfe's works.

Tables

$400

3107 Table
Height 29 inches
Top 36 inches in diameter
Code Word—PUSUP

3107 Table—A reproduction of an antique Duncan Phyfe reading table with revolving top.

$475

2831 Table
Height 29 inches
Top 60 inches x 32 inches
Code Word—OGAUL

2831 Table—A remarkably fine Duncan Phyfe table. Suitable for use in apartment, dining and living room.

Tables

1748 Table
Height 31 inches
Closed 52 x 24 inches
Open 72 x 24 inches
Code Word—AMLUR

1748 Table—A modern interpretation in solid mahogany, of the art of Duncan Phyfe.

2898 Table
Height 30 inches
Top 36 inches x 22 inches
Top 58 inches open
Code Word—PURNI

3528 Table
Height 29 inches
Top 54 inches x 20 inches
Folded 34 x 20 inches
Code Word—PECDU

2898 Table—Duncan Phyfe drop leaf table characteristic of that great cabinet maker's best work.

3528 Table—A simplified model of one of Duncan Phyfe's best examples of the lyre table.

Tables

$425

2674 Table
Height 30 inches
Top 60 inches x 25 inches
Code Word—BOJAX

2674 Table—Ideally suited for a library is this charming adaptation of a 17th Century Table now in the Museum of Industrial Arts, Madrid.

$400

3122 Table
Height 30 inches
Top 72 inches x 32 inches
Code Word—PUSAW

3122 Table—Jacobean table, unusual bulb turned legs, richly carved.

Tables

$370

3124 Table
Height 30 inches
Top 72 inches x 28 inches
Code Word—PUTUR

3124 Table—Simple Jacobean table with incised rail.

$360

3189 Table
Height 30 inches
Top 68 inches x 30 inches
Code Word—PEYOB

3189 Table—Early English design. Most unusual open twist legs. Made in oak; also made in walnut.

Desks

$295

3516 Desk

Height 29 inches   Width 20 inches   Width Open 40 inches   Depth 18 inches
Code Word—PECYO

3516 Desk—A useful and attractive decorated maple desk for small apartments. The top is hinged on the sides, uncovering the letter compartment and extending the writing bed when open.

$325

3518 Desk

Height 40 inches   Width 29 inches   Depth 14 inches   Depth Open 20 inches
Code Word—PEDAS

3518 Desk—A serviceable American decorated maple desk of Sheraton inspiration. Writing bed slides forward for increased writing surface. Decorated doors conceal the letter compartments.

Cabinets and Desks

$500

1767 Desk
Height 58 inches
Width 32 inches
Depth 18 inches
Code Word—AMRIL

1767 Desk—An early American combination desk and bookcase in selected maple, with hand painted decorations.

1491 Cabinet—An Italian cabinet from an original in the Museo Civico collection in Milan. The decorations in colors are by our own artists, following historical models. Ideal for piano rolls. Shelf may be placed at any height. Drawer underneath.

$520

1491 Cabinet
Height 53 inches
Width 28 inches
Code Word—AJJUM

Cabinets

2865 Cabinet and Mirror—Serpentine front Cabinet of Satinwood and Mahogany. 1 drawer, 2 doors. Sheraton design.

$875

$175

$490

2505 Cabinet Desk
Height 60 inches
Width 32 inches
Depth 19 inches
Mahogany Base, Decorated
Code Word—BICRA

2505 Cabinet Desk—Chinese Chippendale; solid mahogany base; imported English brasses; hand carved; desk interior can be used for radio cabinet. Supplied in various shades of Chinese lacquer. May be had with shelf in place of desk interior.

2865 Cabinet
Height 32 inches
Width 42 inches
Depth 21 inches
Code Word—OHAUM

2865 Mirror
Frame 43 x 23 inches
Glass 30 x 20 inches
Code Word—PRAHI

Hand Decorated Furniture

$400

3523 Cabinet
Height 30 inches
Top 30 inches x 14 inches
Code Word—PEJUR

3523 C—Adam Hepplewhite in design, this half Elliptic commode is decorated in a manner consistent with origin. Model from London.

$370

3160 Table
Height 28 inches
Top 36 inches x 22 inches
Code Word—PEJWU

3160 Table—Sheraton.

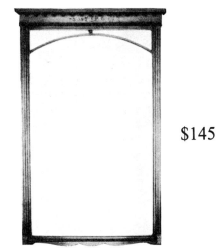

$145

$295

3517 CT
Height 31 inches
Width 39 inches
Depth 15 inches
Code Word—PEKAZ

3517 Mirror
Frame 36 x 24 inches
Glass 32 x 20 inches
Code Word—PEKBA

3517 CT & Mirror—A light and informal maple and decorated console and mirror of American-Sheraton design.

SIMONDS

Consol Suites

2953 CT & Mirror—A Du Barry powder-table with marble top.

1438 CT & Mirror—This exquisite consol table is pure Adam with decorations after the manner of Angelica Kauffman and Pergolisi. Marble or Wood top.

$175

$185

$390

$485

2953 CT
Height 34 inches
Length 32 inches
Width 14 inches
Code Word—BULIM

2953 Mirror
16 inches x 36 inches
Code Word—BULKO

1438 CT
Height 32 inches
Width 43 inches
Depth 16 inches
Code Word—AJEMA

1438 Mirror
24 inches x 39 inches
Code Word—AJERE

Hand Decorated Furniture

3144 Cabinet & Mirror—Venetian School of Italian design. Painted and decorated.

3148 CT & Mirror—French Directoire.

$145

3144 Mirror
34 inches x 22 inches
Code Word—PEJEB

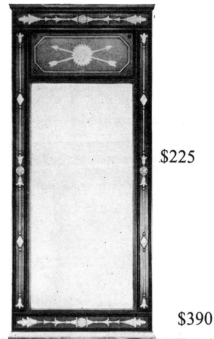

$225

$390

$475

3144 Cabinet
Height 32 inches
Width 32 inches
Depth 15 inches
Code Word—PEJDA

3148 CT
Height 34 inches
Width 39 inches
Depth 18 inches
Code Word—PEJHE

Mirror
49 inches x 21 inches
Code Word—PEJIF

Consol Suites

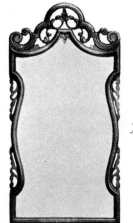

$185

$390

2613 CT
Height 32 inches
Width 47 inches
Depth 17 inches
Code Word—BOHDY

Mirror
Frame
43 inches x 23 inches
Glass
36 inches x 20 inches
Code Word—BOHEZ

2613 Consol Table and Mirror—The period of Charles II in England saw the development of the combination curved and straight lines, so beautifully illustrated in this Consol Table and Mirror.

$200

2542 CT & M—William and Mary table, rich in carved detail and wood combination.

$475

2542 CT & M
Height 36 inches Width 66 inches Depth 18 inches
Code Word—BIMGA

Mirror
46 inches x 27 inches
Code Word—BIMIC

Consol Suites

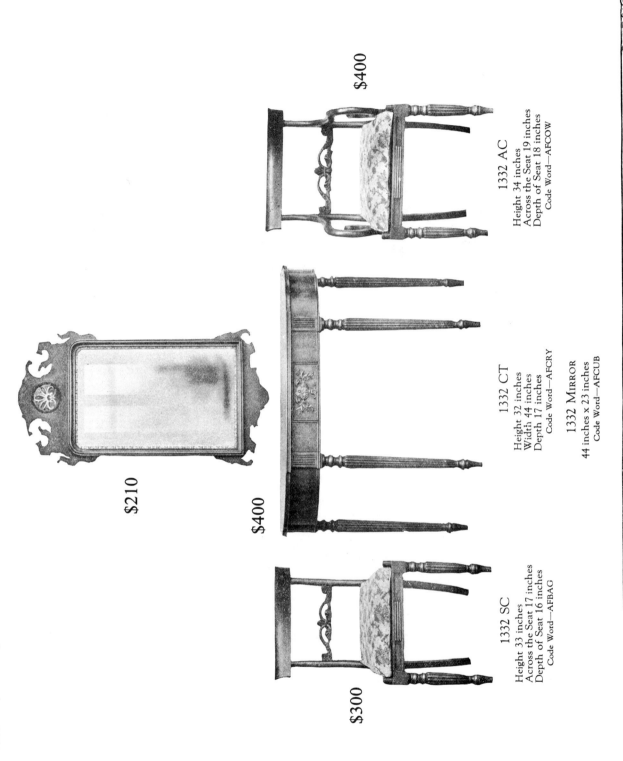

1332 CT, Mirror, SC & AC—This Sheraton Arm Chair was copied from one found at Stratford-on-Avon. Mahogany.

$210

$400

$400

$300

1332 AC
Height 34 inches
Across the Seat 19 inches
Depth of Seat 18 inches
Code Word—AFCOW

1332 CT
Height 32 inches
Width 44 inches
Depth 17 inches
Code Word—AFCRY

1332 Mirror
44 inches x 23 inches
Code Word—AFCUB

1332 SC
Height 33 inches
Across the Seat 17 inches
Depth of Seat 16 inches
Code Word—AFBAG

Breakfast Room Suites

$390 $475 $330

3020 Breakfast Room Suite

TABLE
Open 42 x 49 inches
Closed 42 x 22 inches
Code Word Table—DEBIK

3020 AC
Height 36 inches
Across the Seat 22 inches
Depth 19 inches
Code Word AC—DEBAC

3020 SC
Height 33 inches
Across the Seat 18 inches
Depth 17 inches
Code Word—DEBEG

3020 SUITE—This delightful breakfast room suite is constructed of finest selected birch. It is hand finished. In design, the pieces conform to the French Directoire period, reminiscent of the classic simplicity of Greece, but modified to meet modern needs. Particularly appropriate for display when the first sunny days of Spring create the desire for relief from Winter's formalism.

Breakfast Set

3027 Set

Table	Mirror	Buffet	SC
Height 29 inches	42 inches x 15 inches	Height 36 inches	Height 36 inches
Top 42 inches x 42 inches	Code Word—PARST	Width 60 inches	Across the Seat 18 inches
Code Word—PARTU		Depth 15 inches	Depth of Seat 15 inches
		Code Word—PAROP	Code Word—PARNO
$300	$160	$450	$200

3027 Set—Different and practical is this breakfast suite in the modern manner. Brass supports are a feature of the table.

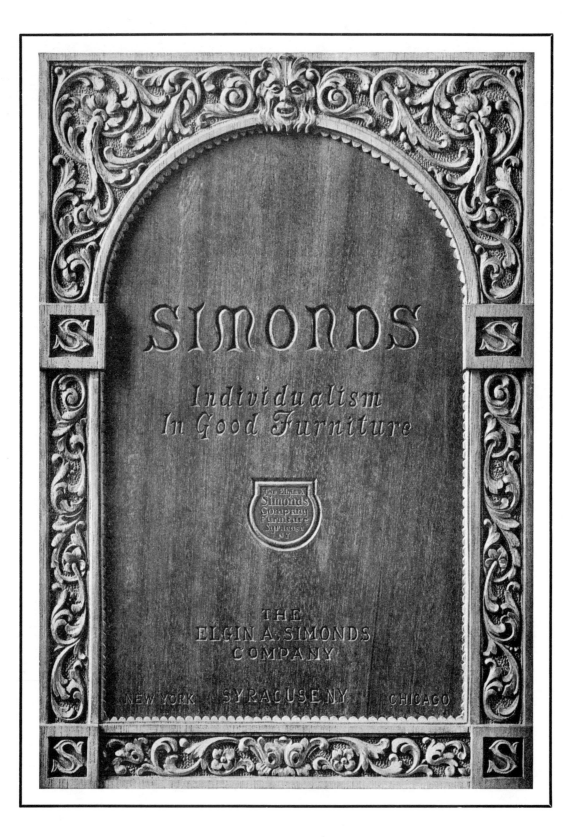

English Novelties

$300

No. 1X69—Coffee Table
Height 13″, top 26″
Low Decorated Coffee Table. English model.

$310

No. 1X75 Coffee Table
Height 14″, top 31″
Walnut Queen Anne Coffee Table—Blue Chinese decorated top.

$300

No. 1X109 Table
Height 28″, top 23″ x 17″

Small size Queen Anne Walnut Table.

Fine in detail. Burl top and walnut banded.

Model from London.

$320

No. 1X56 Nest of Tables
Height 26″, tops 22″ x 15″, 18″ x 13″, 14″ x 12″
Quite unusual is the piecrust edge on the
largest of this Nest of Tables.
Burl Ash tops.

English Novelties

$500

No. 1X114 Folding Top Card Table—Walnut
Height 28″, top 28″ x 28″, folded 28″ x 14″
Beautiful Georgian Folding Top Card Table with accordian extension and leather playing surface decorated. A fine piece of furniture.

$390

No. 1X115—Burl Walnut Top
Height 22″, top 27″ x 21″
A fine Queen Anne Coffee Table. Drawers make this piece very serviceable

$275

No. 1x92 Coffee Table
Height 20″, top 20″ x 15″
Small Queen Anne Coffee Table. Burl Ash Top, Piecrust moulding.

$500

No. 1X114 Folding Top Card Table
Same as No. 1X114 above
Height 28″, top 28″ x 28″, folded 28″ x 14″
Leather decorated cover will be used in place of needlepoint.

English Novelties

$250

No. 1X58 Round Coffee Table—Height 20″ top 22″ dia.

Small size Round Coffee Table. Mahogany fret brackets add greatly to the character of this Chippendale piece.

$275

No. 1X99 Card Table
Height 26″, top 27″ round, folded 27″ x 13″

Small size Queen Anne Card Table. Crotch Mahogany rims—felt top. Can be used as an end table also.

$260

No. 1X60
Height 28″, top 19″ shelf 16″

An amusing but useful Walnut Folding Table. Shelf can be used either one half or full.

Model from England.

$240

No. 1X59
Height 19″, top 22″
A Tip Top Coffee Table—Burl Ash Top.

$240

No. 1X65
Height 20″, top 25″ x 19″

Small Queen Anne Walnut oval Coffee Table. Burl top and banded in walnut.

LENOX

English Novelties

$300

1X67 English Coffee Table
Height 19″, top 27″ round
Curly maple top—Octagonal legs.

$300

No. 1X102
Height 26″, top 21″
Queen Anne Tip Top Table. Butt walnut top with pie crust moulding.

$200

No. 1X54
Height 14″, top 20″
Burl Ash Top with piecrust moulding

$245

No. 1X51
Height 15″, top 30″
English Coffee Table.
Top is of Burl Ash and banded with Circassian Walnut.

English Novelties

$400

No. 1X118 William and Mary Table
Height 30″, top 28″

A wonderful William and Mary Table of Walnut—with decorated mirror top.
Can also be had with wood top. Model from Kensington.

$300

$375

No. 1X119, Wall Table
Height 29″, top 23″ x 13″

An exquisite Queen Anne Wall Table. Burl top and
banded in walnut. Detail in piece exceptionally fine.
Model from Highwycomb.

No. 1X78 Bookcase
Height 35″, width 25″, depth 10″

Walnut Queen Anne Bookcase—Top crotch mahogany
and walnut banded.

LENOX

English Novelties

$300

No. 1X68 Cabinet
Height 29″, top 23″ x 15″

Small size Walnut Cabinet with drawers. Carved moulding on top and drawer—touched off in Gold. Early Queen Anne—Model from Ipswich.

$320

No. 1X96 Lowboy   Height 28½″, top 17″ x 24″
A small Lowboy or Chest of Drawers.

Curly Maple top. Beautiful in every way. Walnut base and decorated top. Model from Highwycomb.

$325

No. 1X61
Height 28″, top 24″ x 16″

Walnut with carved moulding in Top and Drawer—touched off in gold. Burl ash top and banded in walnut. Model from England.

$310

No. 1X123—Drop Leaf Table
Height 29″, top 35″ x 25″, folded 25″ x 18″

A useful Drop Leaf Table—Mahogany Early Chippendale, with drawer.

LENOX

English Novelties

$320

No. 1X63 Table
Height 27″, top 29″ x 17″
Bleached walnut with Chinese Decorated Top.   Model from Sussex.

No. 1X62 Folding Card Table.   Height 28″, top 25″ x 22″, top open 25″ x 25″
Walnut Folding Card Table.   Top turns—small in size.   Burl ash top—carved moulding on top, touched off in gold.   Beautiful in the handling of veneers and proportions;   model from England.

LENOX

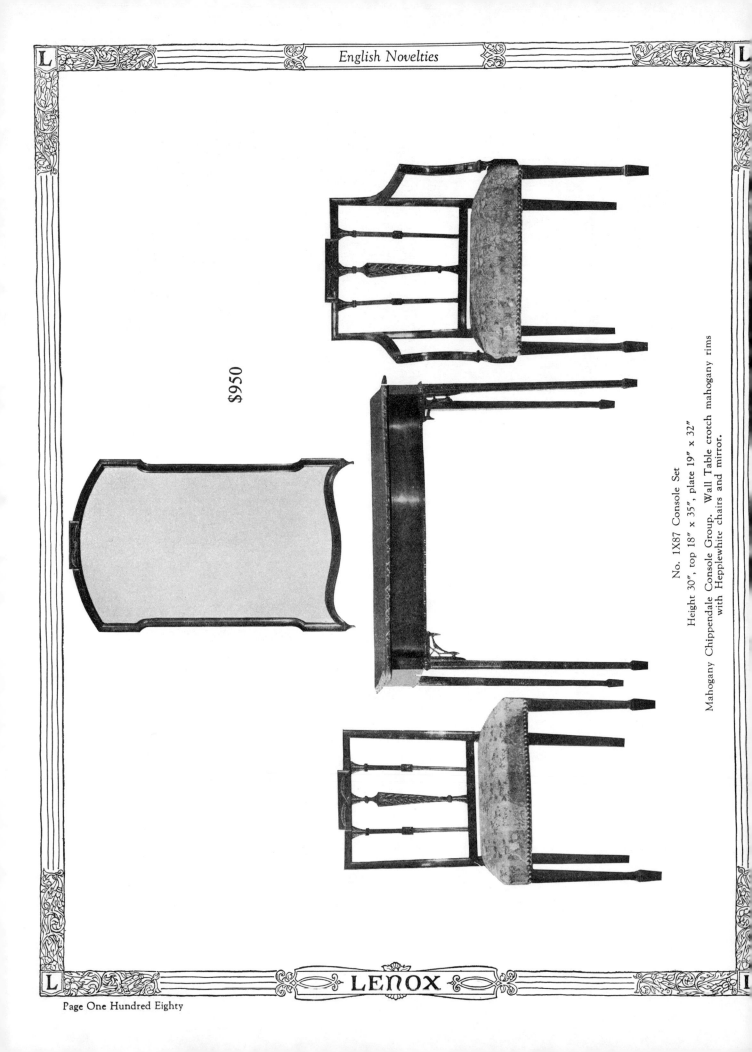

English Novelties

$950

No. 1X87 Console Set
Height 30", top 18" x 35", plate 19" x 32"
Mahogany Chippendale Console Group. Wall Table crotch mahogany rims with Hepplewhite chairs and mirror.

Secretarys

$1200

No. 221 Secretary—Mahogany

An American Secretary of Hepplewhite design with the broken Pendiment, carved shell and small paned doors, typical work of the skilled craftsman.

Automatic supports for desk fall.

No. 221 Secretary
Curly Maple or Mahogany
Height 82", width 36"
depth 19"

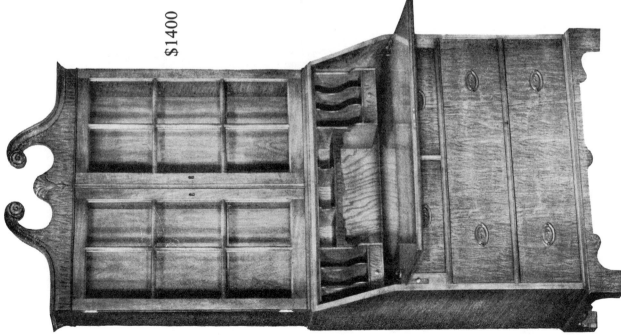

$1400

No. 221 Secretary—Maple

LINCOLN

Desks

$900

A low bureau type desk of the Hepplewhite school as made in the colonies. This desk not only lends a distinctive colonial touch to the home but also serves a useful purpose in providing writing facilities as well as allowing storage in its ample drawers.

No. 221 Desk—Curly Maple
Base 19″ x 36″

$800

No. 221 Desk—Mahogany
Base 19″ x 36″

Highboy Desks

$1600

$1500

No. 227 Highboy
Height 63″, width 36″, depth 20″

Made in Curly Maple or Mahogany.

The graceful cabriole legs with quaint three toed foot, carved fan and delicate Finials are distinct features of the early American Period in the Queen Anne style.

No. 228 Highboy Desk
Height 63″, width 36″, depth 20″

Made in Curly Maple or Mahogany.

The desk drawer makes this decidedly practical.

Lincoln

Bedroom Furniture

$125

No. 696 Bench

$400

No. 696 Bed—4' 6" Slat
No. 696½ Bed—3' 3" Slat
Side Rails 6' 4" long

$400

No. 696½ Bureau
Top 46" x 22"
Hanging Mirror 22" x 30"

$195

No. 696 Straight Chair

LINCOLN

Bedroom Furniture

$250

No. 696 Bedside Table
Top 16″ x 14″

$400

No. 696 Semi-Vanity
Width 42″, depth 19″
Mirrors 8″ x 22″—18″ x 28″

One of the few examples of pure American furniture of 1790-1830. Evidence of colonial simplicity is here shown in adapting Empire Period lines without any of the usual Empire ornamentation.

$500

No. 696 Chiffonier
Top 38″ x 20″

$200

No. 696 Straight Rocker

Scroll Colonial in Mahogany.

LINCOLN

Bedroom Furniture

$400

No. 697 Semi Vanity—Mahogany
Width 42″, depth 19″
Mirrors 8″ x 22″—18″ x 28″

$495

Another example of pure American colonial design. The inspiration for this suite came from the Sheraton school. The elimination of all ornament with the exception of the reeding on posts is characteristic of the period.

No. 697 Bureau—Mahogany
Top 46″ x 22″
Mirror 34″ x 28″

Lincoln

Page One Hundred Eighty-seven

Bedroom Furniture

$300

$195

No. 724 Bedside Table—Maple
Top 16½" x 13½"

No. 724 Bed 4'6" Slat—Maple
724½ Bed 3'3" Slat—Maple
Side Rail 6'4" long

This charming suite is reminiscent of early Pennsylvania Dutch Colonists. Made in Maple, can be had plain or decorated.

$200

$510

$150

No. 724 Rocker—Maple

No. 724 Chiffonier—Maple
Top 36" x 26½"
Movable Toilet Mirror—12" x 16"

LINCOLN

Bedroom Furniture

$125

No. 724 Bench—Maple

$95

$350

No. 724 Toilet-Desk Table
Top 32″ x 19½″
Hanging Mirror 14″ x 22″

$160

$425

No. 724 Bureau
Top 42″ x 20½″
Hanging Mirror 18″ x 28″

$125

No. 724 Chair—Maple

Lincoln

Bedroom Furniture

$190

No. 725 Bedside Table
Top 16″ x 13″

$120

No. 725 Bench—Maple

$160

$150

$400

No. 725 Bureau—Maple
Top 42″ x 20″
Hanging Mirror 18″ x 26″

$400

No. 725¾ Chiffo-Desk—Maple
Top 33″ x 19″
Hanging Mirror 14″ x 22″

No. 725 Chiffonier—Maple
Top 33″ x 19″

$310

LINCOLN

Bedroom Furniture

$140
No. 725 Rocker—Maple

$115
No. 725 Chair—Maple

$300
No. 725 Bed—Maple 4' 6" Slat
No. 725½ Bed—Maple 3' 3" Slat
Side Rail 6' 4" long

$465
No. 725½ Bureau—Maple
Top 42" x 20"
Mirror 22" x 28"

$145

$320
No. 725 Toilet-Desk Table
Top 32" x 19"
Hanging Mirror 14" x 22"

This delightful and popular suite in the Sheraton style will make a place for itself where simplicity and beauty is the keynote.

LINCOLN

Bedroom Furniture

$375

No. 745 Ped 4' 6" Slat
No. 745½ Bed 3' 3" Slat—Side Rail 6' 4"

$210

$425

No. 745 Chiffonier
Top 38" x 20"
Movable Toilet Mirror 14" x 18"

$395

No. 745 Vanity Toilet Table
Width 47", depth 19½"
Mirror 20" x 30"

$125

No. 745 Chair

Made in Curly Maple with matched Satinwood insert.   Also made in Mahogany with Crotch Mahogany insert.

Bedroom Furniture

$425

No. 742 Bed 4' 6" Slat
No. 742½ Bed 3' 3" Slat
Side Rail 6' 4"

$100

No. 742 Bench
Colonial Suite, made in maple or mahogany.

$200

The Mirror with Painting on glass is a reproduction from an original in Independence Hall, Philadelphia. The balance of the suite is of Phyfe-Sheraton design.

$350

No. 742 Desk Table
Top 34" x 20½"

$450

No. 742 Bureau
Top 46" x 21½"
Hanging Mirror 18" x 26"

Bedroom Furniture

$400

No. 742½ Bureau
Top 46" x 21½"
Mirror 24" x 30"

$200

$450

No. 742 Chiffonier
Top 40" x 21½"
Movable Toilet Mirror 14" x 18"

LINCOLN

Bedroom Furniture

$350

No. 747 Chiffonier
Top 30″ x 18″

$200

No. 747 Bedside Table
Top 15″ x 15″

A simple Colonial design in Maple

$340

$100

$310

747½ Toilet Table
Top 30″ x 18″
Mirror 16″ x 20″

No. 747 Toilet-Desk Table
Top 30″ x 18″
Hanging Mirror 14″ x 22″

LINCOLN

Bedroom Furniture

$320

The very pleasing combination of Fiddleback Mahogany, Thuya Burl, Stripe Mahogany Banding on drawers and Hand Painted decorations make this Empire suite wanted where a quiet but imposing dignity is desired.

No. 748 Bed 4' 6" Slat
No. 748½ Bed 3' 3" Slat

No. 748 Chiffonette
Top 40" x 20"
$500

$500

$125

No. 748 Bureau
Top 48" x 21"   Mirror 36" x 28"

No. 748 Chair

LINCOLN

Page Two Hundred Two

Bedroom Furniture

This suite may be decorated in Robins Egg Blue and Hand Decorations. Listed on Price List as No. 756.

$275

No. 749 Bed 4' 6" Slat
No. 749½ Bed 3' 3" Slat
Side Rail 6' 4"

$150

$275

Executed in Macassar Ebony and finely figured Curly Maple, with decoration in silver, the combination of which makes this Art Moderne suite most pleasing.

$150

No. 749 Bureau
Width 50", depth 22"
Hanging Mirror 28" x 28"

No. 749 Bedside Table
Width 17", depth 14"
Height 28"

Bedroom Furniture

$275

No. 751
Bedside Table
Width 16"
Depth 14"

$400

No. 751 Bed 4' 6" Slat
No. 751½ Bed 3' 3" Slat
Side Rail 6' 4"

$150

$410

No. 751 Chiffonier
Top 38" x 20"
Movable Toilet Mirror 14" x 18"

$125

No. 751 Chair

LINCOLN

Bedroom Furniture

Made in Satinwood and Harewood with Hand Painted decorations. Silvered mirrors. This Hepplewhite suite creates a charming effect where daintiness and soft colors prevail.

$500

No. 751 Toilet Table
Top 44" x 17"
Mirror 20" x 26"

$450

$160

No. 751 Bureau
Top 50" x 22"
Hanging Mirror 22" x 32"

No. 751 Bench

LINCOLN

Bedroom Furniture

$475

$220

No. 750 Bedside Table
Top 16″ x 14½″

Queen Anne suite of walnut with finely executed shell carving and beautifully figured woods of the period.

No. 750 Bed 4′ 6″ Slat
No. 750½ Bed 3′ 3″ Slat

Portable Triple Mirror. Frame in jade green with raised gold lacquer decorations.

$100

$480

$380

No. 750½—Consol Triple
Mirror—Top 38″ x 18½″
Mirrors 8″ x 16″—16″ x 20″

No. 750 Chiffonier
Top 40″ x 20″
Movable Toilet Mirror 14″ x 18″

LINCOLN

Bedroom Furniture

$550

$145

$500

No. 750 Toilet Table
Top 44″ x 19″
Mirror 18″ x 28″

No. 750¾ Consol Base
Top 38″ x 18½″
Hanging Mirror 18″ x 28″

Bedroom Furniture

$125

$125

No. 759 Bench
Height 18"
Top 20" x 14"

No. 759 Side Chair
Height 36"
Width of Seat 17"; Depth of Seat 15"

$400

No. 759 Toilet Table
Top 42" x 19"
Glass 20" x 30"

LINCOLN

Bedroom Furniture

$475

No. 759 Chiffonier
Top 36" x 20½"
Movable Toilet Mirror 14" x 18"

$100

$375

No. 759½ Bureau
Top 46" x 20½"
Glass 22" x 32"
and hanging mirror

LINCOLN

Bedroom Furniture

$200

No. 759 Night Stand
Top 16″ x 14″

$495

No. 759 Bureau
Top 46″ x 20½″
Glass 22″ x 32″

$310

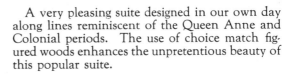

No. 759 Bed—4′ 6″ Slat
No. 759½ Bed—3′ 3″ Slat
6′ 4″ Side Rails

A very pleasing suite designed in our own day along lines reminiscent of the Queen Anne and Colonial periods. The use of choice match figured woods enhances the unpretentious beauty of this popular suite.

Solid Walnut with figured butt veneers.

LINCOLN

Bedroom Furniture

The characteristic Empire motifs glorifying war and the peaceful contrast of graceful curves and beautifully matched mahogany blend happily in this Empire suite.

Mottled Cherry and Mahogany.

$400

No. 753 Chiffonier
Top 33″ x 19″
Plate 12″ x 20″

$390

No. 753 Bureau
Top 44″ x 20″
Plate 16″ x 34″

$300

No. 753 Bed—4′ 6″ Slat
No. 753½ Bed—3′ 3″ Slat

$200

No. 753 Night Stand
Closed
Width 12″ Depth 18″

LINCOLN

Bedroom Furniture

$145

No. 753 Night Stand
Open
Width 12"
Depth 18"

$110

No. 753 Bench

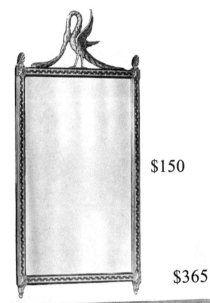

$150

$110

No. 753 Side Chair

$365

No. 753 Toilet-Desk
Top 40" x 18"
Plate 16" x 26"
and hanging mirror

Bedroom Furniture

The charm of this Hepplewhite suite lies in the inlaying and veneering of the finest mahogany and satinwood, together with the beauty of line and form of the designs. Mottled Mahogany and satinwood inlays.

$145

No. 754 Side Chair

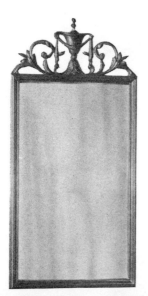

$360

No. 754 Toilet-Desk
Top 40″ x 18¼″
Glass 17″ x 28″
and hanging mirror

$295

$120

No. 754 Bed—4′ 6″ Slats
No. 754½ Bed—3′ 3″ Slats
Bolt Siderail 69″ x 5″ x 1⅛″

Bedroom Furniture

$500

No. 754 Chiffonier
Top 34″ x 21″
Glass 14″ x 18″
and movable toilet mirror

$210

No. 754 Night Stand

$225

$460

No. 754 Bureau
Top 42″ x 22″
Glass 18″ x 30″
and hanging mirror

Lincoln

Bedroom Furniture

A Colonial suite in which the Dutch feeling is evident. This suite is shown here in curly maple. Can be had in Mahogany also.

$100

$490

No. 757 Chiffonier
Top 36" x 20½"
Movable Toilet—12" x 16"

$200

$445

No. 757½ Bureau
Top 44" x 20½"
Glass 20" x 30"
and hanging mirror

$325

No. 757 Bed—4' 6" Slat
No. 757½—3' 3" Slat
6' 4" Side Rails

$175

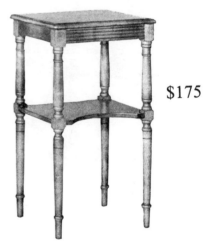

757 Night Stand—Top 16" x 13"

Lincoln

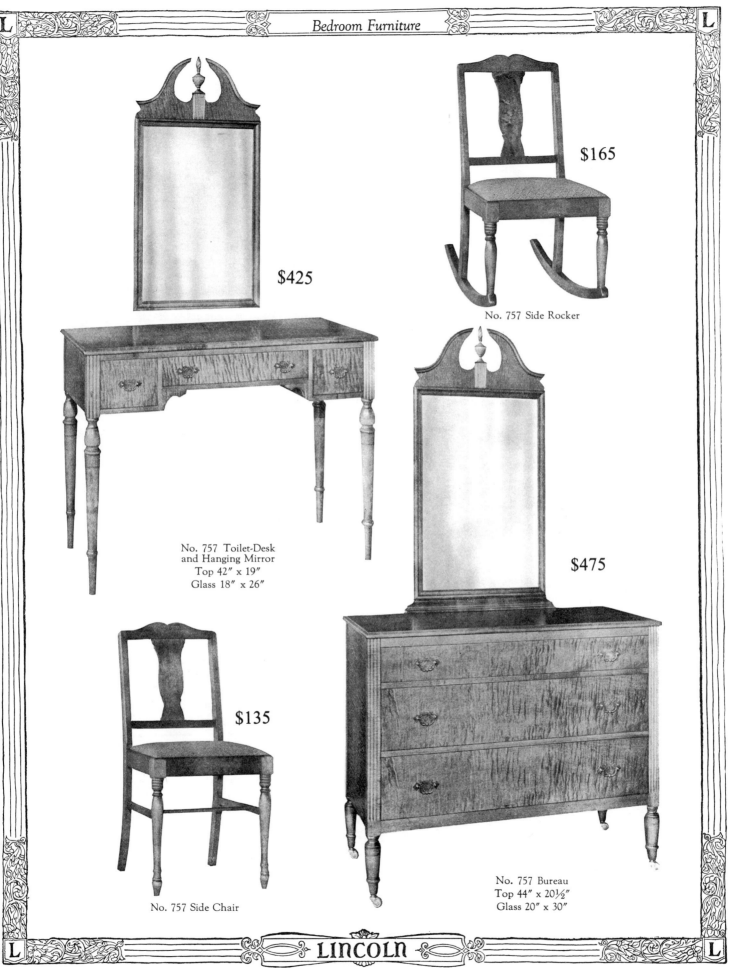

Bedroom Furniture

$425

$165
No. 757 Side Rocker

No. 757 Toilet-Desk and Hanging Mirror
Top 42" x 19"
Glass 18" x 26"

$475

$135

No. 757 Side Chair

No. 757 Bureau
Top 44" x 20½"
Glass 20" x 30"

LINCOLN

INDEX OF LINCOLN FURNITURE

Pattern		Page
221	Secretary	181
221	Desk	182
227	Highboy	183
228	Highboy Desk	183
696	Bench, Bed, Chair	184
696	Semi-Vanity, Bedside Table, Chiffonier, Straight Rocker	185
696½	Bed, Bureau and Hanging Mirror	184
697	Bed, Bedside Table, Chiffonier and Movable Toilet Mirror	186
697	Semi-Vanity, Bureau	187
697½	Bed, Bureau and Hanging Mirror	186
724	Bed, Bedside Table, Rocker, Chiffonier and Movable Toilet Mirror	188
724	Bench, Bureau and Hanging Mirror, Toilet-Desk Table and Hanging Mirror, Chair	189
724½	Bed	188
725	Bedside Table, Bench, Bureau and Hanging Mirror, Chiffonier	190
725	Rocker, Chair, Bed, Toilet-Desk Table and Hanging Mirror	191
725½	Bed, Bureau	191
725¾	Chiffo Desk and Hanging Mirror	190
735	Bench, Toilet Table, Bedside Table, Chair, Chiffonier and Movable Toilet Mirror	192
735	Bed, Bureau, Desk Table	193
735¼	Toilet Table and Hanging Mirror	193
735½	Bed	193
735¾	Bureau and Hanging Mirror	192
742	Bed, Bench, Bureau and Hanging Mirror, Desk Table	196
742	Consol Table and Hanging Mirror, Bedside Table, Chair, Semi-Vanity Toilet	197
742	Chiffonier and Movable Toilet Mirror	198
742½	Bed	196
742½	Bureau	198
745	Bed, Chiffonier and Movable Toilet Mirror, Chair, Vanity Toilet Table	194
745	Toilet Bench, Bureau, Night Stand	195
745½	Bed	194
745½	Bureau and Hanging Mirror	195
747	Chiffonier, Bedside Table, Toilet-Desk Table and Hanging Mirror	199
747	Rocker, Bed	200
747	Bench, Chair, Bureau and Hanging Mirror, Toilet-Desk Table	201
747¼	Bed	201
747½	Toilet Table	199
747½	Bed, Bureau with Swinging Mirror, Chiffonier	200
747¾	Bed	201
748	Bed, Bureau, Chiffonette, Chair	202
748	Toilet Table and Hanging Mirror, Bench, Bedside Table	203
748½	Bed	202
748½	Bureau and Hanging Mirror	203
749	Bed, Bedside Table, Bureau and Hanging Mirror	204
749	Bench, Chiffonier and Movable Toilet Mirror, Toilet Table and Hanging Mirror, Chair	205
749½	Bed	204
750	Bed, Bedside Table, Chiffonier and Movable Toilet Mirror	208
750	Bench, Bureau and Hanging Mirror, Chair	209
750	Toilet Table	210
750¼	Toilet Table and Hanging Mirror	209
750½	Bed, Consol, Triple Mirror	208
750¾	Consol Base and Hanging Mirror	210
751	Bedside Table, Bed, Chiffonier and Movable Toilet Mirror, Chair	206
751	Bureau and Hanging Mirror, Bench, Toilet Table	207
751½	Bed	206
753	Bureau and Hanging Mirror, Bed, Chiffonier and Movable Toilet Mirror, Night Stand	214
753	Bench, Night Stand, Side Chair, Toilet-Desk Table and Hanging Mirror	215
753½	Bed	214
754	Bed, Bench, Side Chair, Toilet-Desk Table and Hanging Mirror	216
754	Bureau and Hanging Mirror, Chiffonier and Movable Toilet Mirror, Night Stand	217
754½	Bed	216
757	Bed, Chiffonier and Movable Toilet Mirror, Night Stand	218
757	Bureau, Side Chair, Side Rocker, Toilet-Desk Table and Hanging Mirror	219
757½	Bed, Bureau and Hanging Mirror	218
759	Bench, Side Chair, Toilet Table	211
759	Chiffonier and Movable Toilet Mirror	212
759	Bureau, Bed, Night Stand	213
759½	Bureau and Hanging Mirror	212
759½	Bed	213

Index

INDEX OF SIMONDS FURNITURE

Pattern		Page	Pattern		Page	Pattern		Page
735	Side Chair	14	1694	Side Chair	23	2509	Smoking Stand	140
735	Arm Chair	14	1694	Arm Chair	23	2519½	Arm Chair	60
			1697	Side Chair	16	2523	Side Chair	24
967	Arm Chair	8	1697	Arm Chair	16	2523	Arm Chair	24
979	Side Chair	5				2528	Sofa	120
981	Arm Chair	9	1702	Easy Chair	92	2531	Arm Chair	54
993	Arm Chair	4	1702	Sofa	92	2541	Side Chair	14
996	Side Chair	4	1704	Arm Chair	68	2541	Arm Chair	14
			1704	Ottoman	68	2542	Console Table	168
1032	Arm Chair	55	1715	Arm Chair	60	2542	Mirror	168
1032½	Arm Chair	55	1718	Side Chair	22	2543	Wing Chair	93
			1718	Arm Chair	22	2548	Arm Chair	52
1125½	Side Chair	21	1718	Settee	22	2555	Side Chair	23
1125½	Arm Chair	21	1741	Side Chair	78	2555	Arm Chair	23
			1741	Arm Chair	78	2555	Love Seat	22
1206	Side Chair	32	1741	Sofa	78	2555	Gate Leg Table	136
1206	Arm Chair	32	1741	Bench	131	2556	Tip Top Table	148
1219	Side Chair	34	1742	Sofa	104	2558	Arm Chair	115
1219	Arm Chair	34	1744	Love Seat	97	2580	Drop Leaf Table	141
1284	Side Chair	37	1746	Octagon Table	152	2595	Arm Chair	60
1284	Arm Chair	37	1746¼	Library Table	152	2596	Arm Chair	55
1290	Side Chair	41	1746½	Library Table	152			
1290	Arm Chair	41	1748	Library Table	159	2602	Smoking Stand	141
1291	Side Chair	37	1760	Side Chair	57	2603	Arm Chair	55
1291	Arm Chair	37	1760	Arm Chair	57	2606	Arm Chair	75
1295	Side Chair	34	1760½	Arm Chair	57	2608	Arm Chair	61
1295	Arm Chair	34	1760	Swivel Side Chair	127	2613	Side Chair	20
			1760½	Swivel Arm Chair	127	2613	Arm Chair	20
1332	Side Chair	169	1760	Love Seat	126	2613	Consol Table	168
1332	Arm Chair	169	1760	Ottoman	57	2613	Mirror	168
1332	Consol Table	169	1766	Side Chair	20	2621	Side Chair	24
1332	Mirror	169	1766	Arm Chair	20	2621	Arm Chair	24
1332	Drop Leaf Table	157	1767	Desk	163	2625	Octagon Table	152
			1773	Wing Chair	107	2626	Sofa	100
1438	Consol Table	166	1777	Arm Chair	67	2627	Arm Chair	60
1438	Mirror	166	1790	Side Chair	17	2628	Cabinet	149
1480	Arm Chair	115	1790	Arm Chair	17	2632	Book Stand	141
1491	Cabinet	163	1795	Arm Chair	68	2634	Love Seat	80
1497	Side Chair	31				2636	Arm Chair	52
1497	Arm Chair	31	1802	Side Chair	8	2637	Arm Chair	51
1497	Bench	132	1802	Arm Chair	8	2639	Arm Chair	79
1497	Smoking Stand	151	1803	Arm Chair	10	2644	Arm Chair	56
1497½	Library Table	151	1803½	Arm Chair	10	2649	Arm Chair	99
			1812	Arm Chair	12	2649	Sofa	99
1509	Arm Chair	56	1824½	Swivel Arm Chair	9	2649½	Arm Chair	98
1567	Arm Chair	74	1832	Arm Chair	8	2649½	Sofa	98
1568	Arm Chair	95	1834	Arm Chair	9	2659	Easy Chair	96
1568	Sofa	95	1836	Arm Chair	12	2659	Wing Chair	96
1568½	Arm Chair	94	1840	Arm Chair	10	2659	Sofa	96
1568½	Love Seat	94	1842	Arm Chair	10	2659	Love Seat	97
1568½	Sofa	94	1850	Arm Chair	12	2668	Wing Chair	71
1576	Side Chair	27	1850	Settee	12	2668½	Wing Chair	71
1576	Arm Chair	27	1858	Arm Chair	5	2674	Table	160
1579	Side Chair	5	1864	Arm Chair	9	2690	Bench	133
1579	Arm Chair	5	1872	Arm Chair	7	2692	End Table	140
1581	Bench	133	1874	Arm Chair	7			
1588	Side Chair	21	1877	Arm Chair	11	2701	Coffee Table	144
1588	Arm Chair	21	1880	Arm Chair	7	2703	Coffee Table	144
			1882	Arm Chair	7	2724	Arm Chair	52
1603	Bench	132	1884	Side Chair	6	2729	Gate Leg Table	149
1605	Bench	133	1884	Arm Chair	6	2760	Arm Chair	61
1606	Wing Chair	75	1886	Arm Chair	6	2768	Wing Chair	71
1617½	Arm Chair	74	1889	Arm Chair	11	2771	Wing Chair	75
1618	Wing Chair	72	1890	Arm Chair	6	2778	Swivel Arm Chair	124
1618	Ottoman	68	1892	Arm Chair	11	2782	Side Chair	13
1641	Arm Chair	51				2782	Arm Chair	13
1642	Arm Chair	49	2169	Side Chair	4	2783	Wing Chair	70
1651	Arm Chair	69	2169	Arm Chair	4	2785	Arm Chair	48
1651	Ottoman	69				2786	Coffee Table	146
1670	Arm Chair	89	2238	Side Chair	27	2792	Arm Chair	56
1670	Sofa	89				2792½	Arm Chair	56
1671	Arm Chair	69	2500	Wing Chair	108	2797	Arm Chair	48
1671	Ottoman	69	2500	Sofa	108			
1672	Arm Chair	77	2502	Side Chair	13	2806	Arm Chair	68
1681	Easy Chair	93	2502	Arm Chair	13	2808	Arm Chair	51
1681	Sofa	93	2505	Cabinet	164			
1684	End Table	149						

Index

INDEX OF SIMONDS FURNITURE—Cont'd

Pattern		Page	Pattern		Page	Pattern		Page
2812	Smoking Stand	140	2982½	Arm Chair	19	3121	Table	155
2814	Dressing Table	119	2982½	Sofa	19	3122	Table	160
2817	Table	153				3123	Table	155
2823	Arm Chair	107	3020	Side Chair	170	3124	Table	161
2823	Sofa	107	3020	Arm Chair	170	3129	Table	146
2824	Table	150	3020	Breakfast Table	170	3130	Stool	142
2825	Arm Chair	54	3027	Side Chair	171	3131	Stool	130
2825½	Arm Chair	54	3027	Octagon Table	171	3134	Table	154
2826	Tub Chair	99	3027	Buffet	171	3135	Table	154
2826½	Tub Chair	98	3027	Mirror	171	3136	Table	157
2828	Love Seat	120	3027	Back Curtain	171	3141	Gate Leg Table	156
2831	Library Table	158	3027	Table Scarf	171	3144	Cabinet	167
2833	Side Chair	36	3027	Mirror	171	3144	Mirror	167
2833	Arm Chair	36	3028	Side Chair	44	3145	Chest	139
2840	Coffee Table	146	3028	Arm Chair	44	3148	Consol Table	167
2844	Arm Chair	78	3028	Table	44	3148	Mirror	167
2845	Arm Chair	49	3028	Sideboard	44	3157	Desk	128
2847	Round Table	153	3028	Crt. Cabinet	45	3158	Revolving Book Cabinet	145
2848	Octagon Table	153	3028	Serving Table	45	3160	Table	165
2849	Arm Chair	114	3028	Screen	45	3164	Chest	139
2849	Love Seat	114	3028½	Arm Chair	45	3165	Tip Top Table	148
2849	Sofa	114	3029	Side Chair	31	3166	Tip Top Table	148
2852	Side Chair	25	3029	Arm Chair	31	3169	Dresser	138
2852	Arm Chair	25	3030	Arm Chair	19	3170	Cupboard	137
2852½	Side Chair	26	3031	Side Chair	38	3171	Table	135
2852½	Arm Chair	26	3031	Arm Chair	38	3172	Bench	135
2852½	Settee	26	3032	Side Chair	43	3173	Table	136
2853	Arm Chair	81	3032	Arm Chair	43	3175	Dressing Table	119
2853	Sofa	81	3033	Side Chair	39	3189	Table	161
2854	Smoking Stand	140	3033	Arm Chair	39	3192	Table	149
2855	Wing Chair	72	3034	Side Chair	36	3193	Table	136
2857	Arm Chair	58	3034	Arm Chair	36	3195	Stand	145
2865	Commode	164	3037	Side Chair	40	3198	Stand	147
2865	Mirror	164	3037	Arm Chair	40			
2866	Arm Chair	48	3038	Side Chair	59	3200	Arm Chair	58
2867	Sofa	89	3038	Arm Chair	59	3201	Arm Chair	90
2868	Arm Chair	51	3039	Side Chair	46	3201	Sofa	90
2869	Arm Chair	69	3039	Arm Chair	46	3207	Swivel Arm Chair	124
2870	Arm Chair	48	3041	Arm Chair	29	3211	Arm Chair	67
2871	Arm Chair	69	3044	Side Chair	43	3211	Ottoman	67
2878	Love Seat	92	3044	Arm Chair	43	3213	Arm Chair	79
2880	Revolving Side Chair	124	3046	Side Chair	33	3213	Love Seat	79
2881	Revolving Side Chair	124	3046	Arm Chair	33	3214	Side Chair	25
2882	Arm Chair	49	3047	Side Chair	42	3214	Arm Chair	25
2882½	Arm Chair	49	3047	Arm Chair	42	3215	Arm Chair	112
2887	Bedroom Chair	115	3049	Side Chair	35	3216	Side Chair	28
2889	Wing Chair	70	3049	Arm Chair	35	3216	Arm Chair	28
2898	Drop Leaf Table	159	3051	Side Chair	38	3216	Swivel Side Chair	126
2899	Side Chair	22	3051	Arm Chair	38	3216	Swivel Arm Chair	126
2899	Arm Chair	22	3052	Side Chair	35	3218	Love Seat	80
			3052	Arm Chair	35	3219½	Arm Chair	90
2903	Arm Chair	61	3056	Side Chair	40	3219	Sofa	91
2909	Side Chair	41	3056	Arm Chair	40	3219½	Sofa	91
2909	Arm Chair	41	3058	Side Chair	46	3221	Arm Chair	58
2928½	Arm Chair	64	3058	Arm Chair	46	3223	Arm Chair	113
2933	Arm Chair	53	3060	Side Chair	39	3223½	Arm Chair	113
2936	Arm Chair	62	3060	Arm Chair	39	3225	Wing Chair	70
2939	Arm Chair	72	3064	Side Chair	42	3228	Wing Chair	71
2952	Side Chair	18	3064	Arm Chair	42	3230	Wing Chair	72
2952	Arm Chair	18	3067	Arm Chair	27	3231	Arm Chair	58
2953	Consol Table	166				3232	Side Chair	32
2953	Mirror	166	3100	Table	146	3232	Arm Chair	32
2955	Arm Chair	74	3101	Coffee Table	142	3234	Arm Chair	81
2956	Arm Chair	74	3103	Table	142	3234	Chaise Longue	81
2962	Arm Chair	62	3104	Stand	139	3239	Arm Chair	112
2964	Side Chair	16	3107	Revolving Table	158	3240	Arm Chair	112
2964	Arm Chair	16	3108	Coffee Table	144	3241	Arm Chair	112
2965	Arm Chair	62	3109	Revolving Table	155	3242	Sofa	122
2972	Side Chair	15	3110	Stool	139	3244	Arm Chair	52
2972	Arm Chair	15	3112	Stool	139	3246	Wing Chair	115
2972	Settee	15	3113	Octagon Table	155	3247	Arm Chair	67
2972½	Side Chair	15	3116	Dressing Table	119	3248	Arm Chair	109
2972½	Arm Chair	15	3118	Table	151	3249	Sofa	109
2972½	Settee	15	3119	Table	153	3251	Easy Chair	103
2982½	Side Chair	19	3120	Drop Leaf Table	142	3251	Sofa	103

Page Two Hundred Twenty-two

Index

INDEX OF SIMONDS FURNITURE—Cont'd

Pattern		Page	Pattern		Page	Pattern		Page
3255	Arm Chair	104	3334	Love Seat	76	3507	Table	147
3255½	Arm Chair	104	3335	Arm Chair	83	3512	Table	150
3256	Side Chair	33	3335	Love Seat	83	3513	Stand	148
3256	Arm Chair	33	3335	Sofa	83	3514	Stand	144
3258	Arm Chair	127	3336	Arm Chair	29	3515	Table	143
3258½	Swivel Arm Chair	127	3337	Wing Chair	82	3516	Desk	162
3268	Arm Chair	76	3338	Arm Chair	66	3517	Consol Table	165
3271	Arm Chair	30	3339	Arm Chair	86	3517	Mirror	165
3272	Bench	132	3339	Sofa	86	3518	Desk	162
3275	Settee	134	3341	Arm Chair	29	3520	Table	147
3281	Easy Chair	106	3342	Wing Chair	114	3521	Consol Table	143
3281	Love Seat	106	3344	Wing Chair	82	3522	Coffee Table	136
3281	Sofa	106	3345	Arm Chair	129	3523	Cabinet	165
3282	Arm Chair	103	3345	Table	129	3524	Nest Tables	143
3283	Bench	134	3345	Smoking Stand	129	3525	Consol Table	143
3285	Sofa	123	3351	Easy Chair	87	3526	Coffee Table	145
3286	Arm Chair	53	3351	Sofa	87	3528	Table	159
3287½	Swivel Arm Chair	128	3353	Arm Chair	30	3529	Revolving Book Stand	145
3288	Arm Chair	70	3355	Arm Chair	47			
3290½	Arm Chair	128	3356	Arm Chair	29	3800	Arm Chair	10
3292	Arm Chair	61	3357	Arm Chair	11	4001	Side Chair	28
3296	Arm Chair	125	3358	Wing Chair	113	4001	Arm Chair	28
3297½	Swivel Arm Chair	125	3360	Arm Chair	77	4002	Arm Chair	64
3298	Sofa	122	3360	Settee	77	4039	Stool	130
3299	Side Chair	26	3361	Arm Chair	29	4041	Stool	130
			3362	Arm Chair	79	4042	Bench	132
3300	Sofa	121	3363	Wing Chair	73	4046	Arm Chair	50
3301	Arm Chair	87	3364	Arm Chair	86	4047	Side Chair	50
3305	Arm Chair	30	3365	Arm Chair	30	4047	Arm Chair	50
3306	Wing Chair	76	3366	Arm Chair	47	4048	Arm Chair	54
3308	Arm Chair	125	3367	Arm Chair	47	4049	Side Chair	63
3308½	Swivel Arm Chair	125	3369	Arm Chair	53	4049	Arm Chair	63
3309	Sofa	123	3371	Wing Chair	85	4057	Arm Chair	113
3313	Table Chair	65	3373	Sofa	121	4062	Side Chair	18
3314	Wing Chair	73	3375	Side Chair	63	4062	Arm Chair	18
3315	Wing Chair	88	3375	Arm Chair	63	4066	Side Chair	17
3316	Arm Chair	59	3376	Arm Chair	47	4066	Arm Chair	17
3317	Wing Chair	73	3377	Arm Chair	53	4080	Arm Chair	111
3318	Wing Chair	82	3378	Arm Chair	88	4081	Arm Chair	111
3318	Settee	82	3378	Sofa	88	4081	Side Chair	111
3319	Wing Chair	73	3379	Chaise Longue	80	4081	Love Seat	111
3320	Wing Chair	66	3380	Arm Chair	76	4081	Coffee Table	141
3320	Love Seat	66	3381	Arm Chair	110	4092	Stool	130
3321	Arm Chair	15	3382	Arm Chair	102	4097	Arm Chair	50
3324	Arm Chair	65	3382	Sofa	102			
3326	Wing Chair	84	3383	Bench	131	5018	Arm Chair	64
3327	Easy Chair	84	3384	Arm Chair	110	5020	Arm Chair	64
3327	Sofa	84	3385	Arm Chair	110	5020	Reading Table	150
3328	Arm Chair	85	3386	Arm Chair	110	5021	Pedestal	150
3328	Sofa	85	3388	Arm Chair	109	5035	Easy Chair	101
3329	Stool	129				5035	Wing Chair	101
3330	Table Chair	65	3502	Chest	116	5035	Love Seat	100
3330	Table	65	3502	Bureau	117	5035	Sofa	101
3331	Arm Chair	59	3502	Vanity Table	117	5035½	Arm Chair	100
3332	Arm Chair	102	3502	Hanging Mirror	117	5036	Arm Chair	108
3333	Arm Chair	105	3502	Stand	118	5046	Arm Chair	62
3333	Wing Chair	105	3502	Toilet Mirror	118	6005	Arm Chair	95
3333	Sofa	105	3502½	Bed	116	6013	Love Seat	75

INDEX OF LENOX FURNITURE

Pattern		Page	Pattern		Page	Pattern		Page
1 x 51	Coffee Table	176	1 x 65	Coffee Table	175	1 x 99	Card Table	175
1 x 54	Coffee Table	176	1 x 67	Coffee Table	176	1 x 102	Tip Top Table	176
1 x 56	Nest of Tables	173	1 x 68	Cabinet	178	1 x 109	Table	173
1 x 58	Round Coffee Table	175	1 x 69	Coffee Table	173	1 x 114	Folding Card Table	174
1 x 59	Tip Top Coffee Table	175	1 x 75	Coffee Table	174	1 x 115	Coffee Table	174
1 x 60	Folding Table	175	1 x 78	Bookcase	177	1 x 118	William and Mary Table	177
1 x 61	Cabinet	178	1 x 87	Console Set	180	1 x 119	Wall Table	177
1 x 62	Folding Card Table	179	1 x 92	Coffee Table	174	1 x 123	Drop Leaf Table	178
1 x 63	Table	179	1 x 96	Lowboy	178			

Telegraph Orders

Special Instructions Arranged for Your Convenience and Economy in Ordering by Wire

EXAMPLE

A letter or night letter containing the following message: "If you can ship by October 1st, enter our order for one number sixteen seventy sofa, one each fourteen forty-five arm chair and ottoman. (Stop). One easy chair seventeen naught two. We furnish covers. Do not enter order unless you can ship as specified. Wire answer." (47 words).

Could be coded as follows:

    inbye October 1st
    algew ajicu ajime amaco inaze pryon prypo (10 words)

FOR INSTANCE

Wire acknowledgment and enter order if you can ship by................	= inbye
October first................................	= October first
One 1670 sofa walnut......................	= algew
One 1445 arm chair mahogany.............	= ajicu
One 1445 ottoman..........................	= ajime
One easy chair 1702.......................	= amaco
Cover furnished by us......................	= inaze
Do not enter order unless you can ship as specified	= pryon
Wire answer.................................	= prypo

SPECIAL INSTRUCTIONS

Express immediately...	inadi
Ship by R. R. freight..	inajo
Ship by boat freight...	inbov
Wire if and when you can furnish................................	inaty
Quote prices in lots of..	inava
Cover furnished by us...	inaze
Confirming by mail..	inbag
Mail prints of...	inbek
Our order (number) or (dated)..................................	inbio
Wire acknowledgment and enter order if you can ship by........	inbye
Covered as shown in catalog...................................	incel